22. Colloquium
der Gesellschaft für Biologische Chemie
15.–17. April 1971 in Mosbach/Baden

The Dynamic Structure of Cell Membranes

Edited by D. F. Hölzl Wallach and H. Fischer

With 87 Figures

Springer-Verlag Berlin · Heidelberg · New York 1971

DONALD F. HÖLZL WALLACH, M.D., Director of Radiobiology and Professor of Therapeutic Radiology, Tufts-New England Medical Center, Boston, MA 02111/USA

Professor Dr. HERBERT FISCHER, Max-Planck-Institut für Immunbiologie, 7800 Freiburg-Zähringen, Stübeweg 51

ISBN 3-540-05669-6 Springer-Verlag Berlin·Heidelberg·New York
ISBN 0-387-05669-6 Springer-Verlag New York·Heidelberg·Berlin

This work is subject to copyright. All rights are reserved, whether the whole or part of the material is concerned, specifically those of translation, reprinting, re-use of illustrations, broadcasting, reproduction by photocopying machine or similar means, and storage in data banks.

Under § 54 of the German Copyright Law where copies are made for other than private use, a fee is payable to the publisher, the amount of the fee to be determined by agreement with the publisher.

© by Springer-Verlag Berlin·Heidelberg 1971. Library of Congress Catalog Card Number 73-185 882. Printed in Germany. The use of general descriptive names, trade names, trade marks, etc. in this publication, even if the former are not especially identified, is not to be taken as a sign that such names, as understood by the Trade Marks and Merchandise Marks Act, may accordingly be used freely by anyone.

Druck: fotokop, Darmstadt.

Contents

Introduction. H. FISCHER (Freiburg-Zähringen) 1

Molecular Membranology. F. O. SCHMITT (Brookline) 5

Contacts and Communications Between Cells and Their Relationship to Morphogenesis and Differentiation. R. AUERBACH (Madison) 37

Some Aspects of the Dynamics of Cell Surface Antigens. U. HÄMMERLING (New York) 51

Glycolipid Changes Associated with Malignant Transformation. S. HAKOMORI (Seattle) 65

Structure and Biosynthesis of Viral Membranes. H.-D. KLENK (Gießen) 97

Metabolite Carriers in Mitochondrial Membranes: the Ca^{2+} Transport System. A. L. LEHNINGER (Baltimore) 119

Membrane Phospholipid Metabolism During Cell Activation and Differentiation. E. FERBER (Freiburg-Zähringen) 129

Structure and Function of Hydrocarbon Chains in Bacterial Phospholipids. P. OVERATH, H.-U. SCHAIRER, F. F. HILL, I. LAMNEK-HIRSCH (Köln) 149

Some Aspects of the Structure and Assembly of Bacterial Membranes. L. I. ROTHFIELD (Farmington) 165

Cooperativity in Biomembranes. D. F. HÖLZL WALLACH (Boston) 181

Magnetic Resonance Studies of Membranes and Lipids. J. C. METCALFE (Cambridge/GB) 201

Synthetic Lipid- and Lipoprotein Membranes. H. KUHN (Göttingen) 229

Round Table Discussion 235

Acknowledgements

Our special thanks go to Professor Ernst Auhagen, as treasurer of our Society, who proved himself to be a generous and ever-willing helper. Without him and the "Stiftung Volkswagenwerk", the publication of this Colloquium and the subsequent "round-table discussion", would not have been possible.

We also wish to acknowledge the help of numerous colleagues of the Max-Planck-Institut für Immunbiologie during the Symposium and last, but not least, the secretaries, Mrs. Gisela Bliemeister, Miss Sandra DiCarlo, Mrs. Monika von Lübtow and Mrs. Elfriede Mertens, who handled all the correspondence and were obliged to carry out the difficult tasks of typing, translating and correcting the manuscripts.

Finally, we want to thank the Springer-Verlag and particularly Dr. H. Mayer-Kaupp for rapid publication of this volume.

Freiburg, November 1971

D. F. H. WALLACH

H. FISCHER

Introduction

HERBERT FISCHER

Max-Planck-Institut für Immunbiologie, Freiburg-Zähringen

With 3 Figures

Ladies and Gentlemen:

On behalf of the organizers of the 22nd Mosbach Colloquium, Msrs. HÖLZL-WALLACH, STOFFEL, WIEGANDT and myself, I bid you all a hearty welcome.

We thank you all for coming and naturally feel particular appreciation for the presence of the invited speakers. But, thanks to the tradition that the Mosbach Colloquia have enjoyed for 22 years we did not need to work very hard, since most of our invitations were accepted without hesitation.

Perhaps some of you will wonder why Mosbach and its tradition means so much, especially to the older ones amongst us. In any event, at a time when we were much hungrier and thirstier than we are today, Mosbach became a unique place where we could satisfy our spiritual as well as our physical hunger. It was here where we could find the friendly and peaceful atmosphere which helped us to establish contacts with colleagues from foreign countries and from distant scientific fields, which often led to lasting communication and cooperation.

The initiator of these Colloquia, my teacher Kurt Felix, imparted to these gatherings a pioneer spirit which is more needed today than in the past, particularly because we are now 500 rather than 50 to 100 participants. Indeed, we as organizers, have had to ask ourselves whether it is still possible to have an exciting lecture series combined with the leisure and opportunity for spontaneous questioning and stimulating individual discussion. We are optimistic about this in having decided to be highly selective in our choice of topics. In addition, we have chosen to follow a procedure which we hope that you will understand and approve: *firstly*, we have dispensed with short communications and prepared comments even though this has already forced us to reject a considerable number of worthy communications.

Secondly, as an innovation we have reserved the greater part of the last morning for a round-table discussion. We are extremely grateful to Prof. F. O. SCHMITT for his decision to plan this session and to moderate it. We anticipate that this discussion gives us the

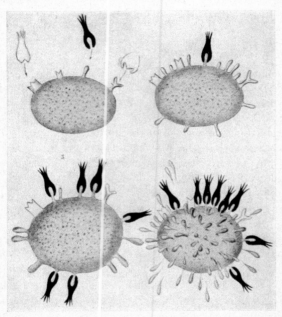

Fig. 1 a—d. *Diagrammatic representation of the side-chain theory*. a "The groups (the haptophore group of the side-chain of the cell and that of the food-stuff or the toxin) must be adapted to one another, *e.g.*, as male and female screw (PASTEUR), or as lock and key (E. FISCHER)." b "... the first stage in the toxic action must be regarded as being the union of the toxin by means of its haptophore group to a special sidechain of the cell protoplasm." c "The side-chain involved, so long as the union lasts, cannot exercise its normal, physiological, nutritive function..." d "We are therefore now concerned with a defect which, according to the principles so ably worked out by...WEIGERT, is...(overcorrected) by regeneration". (From P. EHRLICH [1])

opportunity to focus on some cardinal aspects of present and future membrane research, and also to provide time for a more detailed examination of several aspects which were touched on in the main lectures, but require further clarification.

Finally, I want to comment on our semi-authoritarian emphasis on the dynamic structure of the *plasma membrane* as the major topic of this meeting, thereby leading into the colloquium proper.

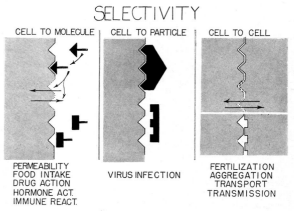

Fig. 2. Diagrammatic representation of selective acceptance (top portion) or nonacceptance (bottom portion) by a cell surface of molecules, particles, or cells of conforming (top) or nonconforming (bottom) configurations, respectively. Arrows indicate the resulting flux across the cell boundary. (From P. WEISS [2])

Quite simply, we are convinced that the study of the plasma membrane is central to innumerable aspects of biomedicine. But we are not the first to arrive at this conclusion: PAUL EHRLICH, whom I, as immunologist, like to consider as one of the fathers of molecular biology, was the first to address this issue in detail, postulating specific receptors for nutrients, regulatory agents, drugs, poisons, and foreign determinants on the cell surface (Fig. 1), which would combine with their specific ligands and thereby modify the surface and the cell [1].

PAUL WEISS formulated EHRLICH's concept in more modern terms, suggesting at the same time that cell surfaces contain diverse domains bearing receptors complementary to specific molecules, particles and even other cells (Fig. 2) and that these surface patterns and specificities change dramatically during differentiation and evolution (Fig. 3) [2]. But despite the obvious validity of these, at one time visionary concepts, and an explosive recognition of additional cell surface processes, we know all too little about membranes and membrane receptors, but we anticipate

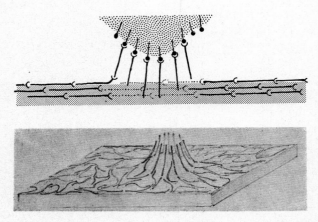

Fig. 3. Diagrammatic representation of the radial reorientation of tangential surface molecules with specific end groups in the vicinity of an external carrier of complementary groups, resulting in a local "microbreach" of the cell boundary. (From P. WEISS [2])

that our horizons will be much wider and our understanding far deeper by the end of this meeting.

References

1. EHRLICH, P.: On immunity with special reference to cell life. In: PAUL EHRLICH: Gesammelte Arbeiten, S. 178—195. (HIMMELWELT, F., Ed.). Berlin-Göttingen-Heidelberg: Springer. London-New York-Paris: Pergamon Press 1957.
2. WEISS, P.: Molecular reorientation as unifying principle underlying cellular selectivity. Proc. nat. Acad. Sci. (Wash.) **46**, 993—1000 (1960).

Molecular Membranology*

F. O. Schmitt

*Neurosciences Research Program,
Massachusetts Institute of Technology,
280 Newton Street, Brookline, MA 02146, USA*

With 12 Figures

It was my assignment in this Colloquium on the Dynamic Structure of Cell Membranes to portray significant recent developments in membranology and to suggest concepts and techniques that merit discussion in the course of the symposium. My role is more that of commentator and catalyst than expert in each of the selected topics. The discussion should also be tutorial, bringing to the attention of chemists certain subjects that are basic in biologic theory and in current experimentation.

My contribution will be heavily slanted toward neuroscience, the study of the physical and chemical substrates of behavior. This science is rapidly moving to the center of interest of the life sciences, indeed of all sciences, because of the pivotal bearing of neuroscience on man's concern about the physical nature and potentialities of his mind and of his very being. Central to neuroscience is membranology, not only because of its role in neuronal excitation and in synaptic function, but also because of the high probability that plasticity, learning, and memory may be somehow imbedded in, and triggered at, the membrane of brain cells, neuronal and glial.

The cell membrane is not merely a static lipid-protein barrier between intra- and extra-cellular milieus, a repository for explicitly localizable molecular machines that pump ions, mediate specific

---

* The Neurosciences Research Program, sponsored by the Massachusetts Institute of Technology, is supported in part by National Institutes of Health Grant No. GM10211, National Aeronautics and Space Administration Grant Nsg 462, Office of Naval Research, The Rogosin Foundation, and Neurosciences Research Foundation.

permeability and bioelectric effects, but is also a highly dynamic, solid state "phase", as suggested by PARDEE [1]. The concept of the membrane as a phase has long been the basis of many valuable contributions by KATCHALSKY [see 2—4] with particular reference to the application of the thermodynamics of irreversible reactions, hysteresis, and the transaction of information in polymeric systems.

It is commonly believed that crucial secrets of brain function will be found in the membrane. Discoveries of membrane organization may well rank among the greatest of molecular biology. But our knowledge of the molecular organization of the membrane, particularly the neuronal membrane, is probably as primitive today as was our knowledge of nucleic acid 30 years ago [5]. We may indeed derive a lesson about membrane structure and function from the history of nucleic acid chemistry. Forty years ago, as a result of extensive purification, nucleic acid was described chemically as a tetranucleotide, C-G-A-T, with molecular weight ca. 1,500. It was difficult to see how this tetranucleotide could play a role in genetics and biosynthesis, as investigators were then suggesting; it seemed more reasonable that the much larger protein molecules should play the central role in genetics. When gentle conditions of extraction were applied, nucleic acid (DNA) was found to be a long, stringy molecule with a very high molecular weight. The basis was thus laid for the discovery that genetic and biosynthetic information is encoded in a giant polymer, DNA, by virtue of a sequential ordering of four codons: adenosine, thymidine, cytosine, and guanosine.

Is it possible that, as in the case of DNA in the mid-thirties, the rather drastic present-day methods of preparation and purification of membranes are masking their true organization and biological role? Currently membranes are isolated by grinding and homogenization of tissues followed by differential centrifugation and "purification", i.e., removal of other organelles and non-membranous material, by various biochemical procedures, some of which may well disrupt vital secondary linkages. Although the resultant material may resemble membranes when examined in the electron microscope after fixation and sectioning of intact tissue, gross alterations may nevertheless have been imposed, particularly on macromolecular and enzymatic complexes loosely bound to the membrane under physiological conditions.

H. Hydén (Gothenburg) has developed a method [personal communication] by which large individual neurons from the Deiters nucleus of the brain may be freed from associated glial cells, cut open with a microknife, freed of nucleus and cytoplasm by gentle saline lavage, and the membrane stretched out on the glass slide, as shown in Fig. 1. Such membranes, which have been subjected

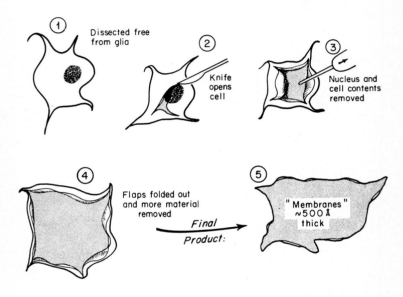

Fig. 1. Method of preparation, by microdissection, of isolated native neuronal membrane. Courtesy of H. Hydén

to no shearing stresses or centrifugation, are found by Hydén to have a thickness of about 500 Å. Their chemical properties may be investigated by picochemical methods developed in Hydén's laboratory over the years. These preparations may also be studied by freeze-etch and scanning electron microscopy. It remains to be seen whether such studies will reveal new facts and will lead to new concepts of the molecular organization and the function of membranes.

The Structure and the Composition of Membranes

The functional interface between intra- and extra-cellular compartments has been called the *greater membrane* [6, 7]. It consists of a thin (ca. 100 Å) *inner zone*, corresponding to the "unit" mem-

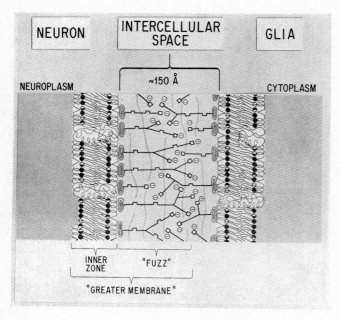

Fig. 2. Diagrammatic representation of molecular organization of the "greater membrane". Lipid bilayer of inner membrane shown as bounded and interpenetrated by protein constituents. Outer layer contains elongate glycoprotein and glycolipid molecules extending into intercellular space. Strongly acidic sialic acid residues shown at chain terminals. Polysaccharides unbonded to membrane shown longitudinally oriented in intercellular space

brane, whose structure is currently much debated, and an *outer zone* or cell coat [8—11]. A possible model of the greater membrane is shown in Fig. 2.

The Inner Zone

The trilaminar hypothesis of DANIELLI and DAVSON [12] was in agreement with conclusions from contemporaneous studies of polarization optics (BEAR and SCHMITT [13]; W. J. SCHMIDT [14]) and X-ray diffraction (SCHMITT et al. [15, 16]). Recent more refined X-ray studies (see R. S. BEAR [17]; DANIELLI and GREEN [18]) and experiments with electron spin resonance (McCONNELL [19]) have supported the lipid bilayer hypothesis of the structure of nerve myelin, which contains lipid and protein in a ratio of about 4:1 by weight. In more characteristic cell membranes, for example, the limiting envelope of erythrocytes and the inner membranes of mitochondria, in which the weight ratio of lipid to protein is near or below unity, it is probable that protein or conjugated protein constituents, at least in some places, may extend through the lipid portion of the membrane. Studies using optical rotatory dispersion, circular dichroism, and nuclear magnetic resonance methods support this possibility (see particularly papers from the laboratories of D. F. H. WALLACH [20, 21] and of S. J. SINGER [22]; see also SCHMITT and SAMSON [7]). With certain qualifications (see URRY [23]), it is possible from such data to derive a rough figure for the relative amount of helical, straight-chain, and random-coil configuration that characterizes the tertiary structure of membrane molecules. However, such evidence can probably not be used in a rigorous manner to prove particular models of membrane ultrastructure (see also the excellent review by ROTHFIELD and FINKELSTEIN [24]). Contributions of electron spin labeling and nuclear magnetic resonance to a study of membrane structure have been made by METCALFE [25], who summarizes the details of his findings in this symposium.

The concept that the typical membrane contains a monolayer of globular "structure protein" molecules, in favor a decade ago, is no longer generally believed, although there is some evidence supporting the notion that the membrane includes globular proteins "intrinsic" to the inner membrane, as well as "extrinsic" proteins, i.e., enzyme clusters, pumps, etc. (see DANIELLI and GREEN [18]). McCONNELL [19], using electron spin labeling, concludes that ca. 25% of the membrane has relatively high fluidity, and he suggests that permeases might function in such regions (McCONNELL's model is shown in Fig. 3). A high degree of unsaturation of some of the lipid aliphatic chains, such as has recently been found to characterize the lipids of the axonal membrane (S. FISCHER, personal communication, 1970; FISCHER et al. [26]), favors fluidity, while complexing of the lipid with surface proteins, glycoproteins, or glycolipids favors solidity of membrane texture. More recent studies of HUBBELL and McCONNELL [27] propose that the lipid fatty acid chains may be relatively stiff rods up to about 8 carbon atoms away from the glycerol polar end, but that more distally, toward the CH_3 end, the chains may be more freely flexible and fluid, depending on degree of unsaturation and other factors. McFARLAND and McCon-

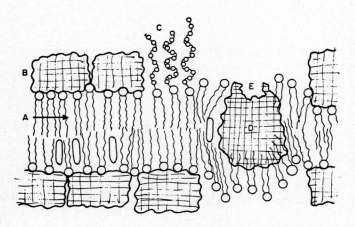

Fig. 3. Schematic and speculative diagram of a membrane structure, showing lipophilic protein D with polar lactose binding site E, to provide facilitated transport of lactose through the fluid hydrophobic region. A. Phospholipids arranged in a bilayer. B. Protein. C. Lipopolysaccharides. D. Permease protein. E. Binding site. From McCONNELL [19]

NELL [28] suggest that the aliphatic chains are kinked; after about the 8th carbon atom the chains are bent at an angle of about 30° (see Fig. 4).

KORNBERG and McCONNELL [29] have made an interesting study of the "flip-flop" action by which a lipid may be transposed with polar end on one side of a bilayer to rotate its orientation through 180° to take up a position with polar ends on the opposite side of the bilayer (Fig. 5). However, the kinetics of this transposition suggests that it is not rapid enough to be an important mechanism for transport through the membrane of ions, bound to the polar head. TRÄUBLE [30] has studied the movement of molecules across membranes in terms of thermal fluctuations of the lipid hydrocarbon chains. Evidence is adduced for the presence of conformational "kink-isomers" which are depicted as providing mobile free volumes in the hydrocarbon phase into which small molecules may

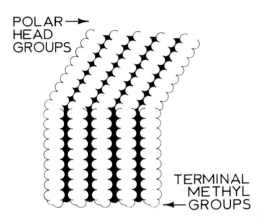

Fig. 4. Schematic representation of the packing of fatty acid chains in one half of a planar lecithin bilayer. Greater motional freedom at ends of methylene chains at terminal groups results from this packing. From McFARLAND and McCONNELL [28]

enter and be transported across the membrane. Calculation of diffusion coefficients suggests that such diffusion is a fast process and may play a significant role in membrane permeation by small molecules.

He also made another recent significant contribution to molecular membranology in his detailed study of the properties of vesicles and aqueous dispersions of pure and synthetic lipids. As shown by CHAPMAN and collaborators [31—34], lipids undergo reversible and endothermic change in the transition, which extends over more than ca. 6 °C, from crystalline to the paracrystalline state. TRÄUBLE studied transition kinetics with temperature-jump method and observed relaxation of ANS fluorescence and changes of optical density of dyes and of light scattering. He concludes that there is cooperative interaction between lipid molecules at transition points and hysteresis in the temperature cycle between metastable states. The nature of the hysteresis depends on the detailed specifications of the lipid molecules and on the presence of non-lipids. If such hysteretic systems, like their ferromagnetic analogues, are to be effective in storing and processing of information, as suggested by KATCHALSKY and OPLATKA [2], the hysteretic cycle must be short (ca. msec.).

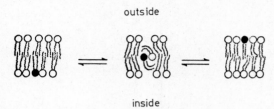

Fig. 5. Schematic representation of "flip-flop" transposition of phospholipid molecules within bilayer. From KORNBERG and McCONNELL [29]

The types of lipids, their polar groups, chain lengths, degree of unsaturation, etc., TRÄUBLE points out, are probably specific for each cell type, possibly for each individual cell, and these factors are reflected in the physiological behavior of the cell. Lipids, therefore, are not indifferent structural components of membranes, but participate in the chemical determination of membrane properties — a point of view that was long ago advocated by LEHNINGER [35].

Nerve myelin is laid down upon the outgrowing nerve axon by satellite cells — Schwann cells in peripheral nerve, glial cells in the central nervous system — and provides insulating material which, interrupted at nodal points separated by ca. 1 mm (representing the ends of each satellite cell), makes possible fast conduction of the action wave by saltatory flow of current from node to node. GEREN [36] showed that the satellite cell produces myelin by wrapping its plasma membrane many times around the axon and extruding cytoplasmic material from the space between the helical wrappings to form compact repeating layers of membrane (Figs. 6 and 7). The loosely packed early myelin ("premyelin") (see AGRA-WAL et al. [37]) is converted into mature myelin by a number of chemical processes including the addition of a basic protein, that bonds electrostatically to the acidic lipids, and the synthesis and the incorporation into the membrane of the long galactolipids or

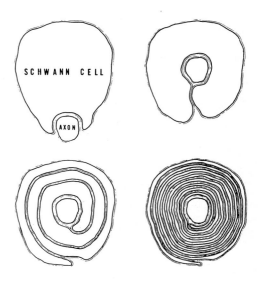

Fig. 6. Schematic representation of the membrane theory of the origin of nerve myelin (after GEREN [36]). Four stages in the engulfment of the axon by the SCHWANN cell and the wrapping of many double membranes which, after condensation, form compact myelin

cerebrosides [38—40] which, possibly because of their pentose groups, favor membrane adhesion.

Early X-ray diffraction results in the writer's laboratory [15, 16] showed that the radial repeating period of 170 to 190 Å included two lipid bilayer-containing membranes and that the lipid bilayers of each membrane contained a mixture of lipids (chiefly phospholipids, galactolipids, and sterols) in smectic, paracrystalline phases. Further analysis of the X-ray data was greatly aided by the discovery that when myelinated nerve fibers are placed in hypoosmotic media, water is intercalated between the cylindrical wrappings of double membranes giving rise to a much larger (> 300 Å) radial repeat resulting from the regular addition of water [41—43]. The membrane pairs, adhering at the surface facing the original satellite cell cytoplasm, exemplify the asymmetric structure of the membranes; i.e., one side facing satellite cell cytoplasm and the

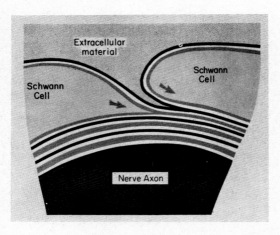

Fig. 7. Illustrating the enfolding of SCHWANN cell surface membrane to form compact myelin. Asymmetry of membrane corresponding to differing chemical composition on external and cytoplasmic sides is shown by dark and light shadings, respectively

other side facing extracellular space (see BLAUROCK and WORTHINGTON [44].

The situation described makes possible assignment of signs for the phases of the small-angle diffractions and the construction of an electron density profile (for a detailed description and evaluation of recent work, see R. S. BEAR [17]). This evidence is helpful in determining the location of substances, such as the basic protein (on the cytoplasmic side of each membrane pair), which have been discovered and characterized by analytical neurochemists. In the current work of CASPAR and KIRSCHNER [45], asymmetries in the electron density profile are interpreted to indicate that cholesterol is present in higher concentration on the side of the membrane pairs facing the extracellular space (see Fig. 8). From the recent

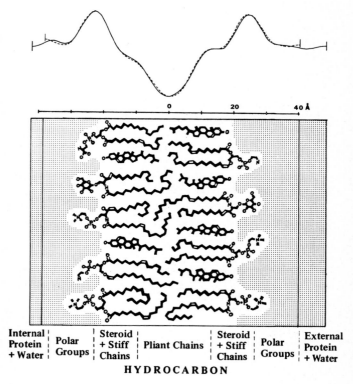

Fig. 8. Schematic illustration of nerve myelin membrane structure. Density profile shown above. Asymmetry of membrane, indicated by difference in steroid content on external, as compared with internal, side of the double membrane. From CASPAR and KIRSCHNER [45]

renewed attack by X-ray and neutron diffraction methods in a number of laboratories will come a more detailed description of the disposition of lipids and proteins in this specialized membrane system. The results will be supplemented by those derived from freeze-etch methods [46, 47], which give high resolution pictures of structure along single individual membranes, rather than average information characteristic of physicochemical methods.

The Outer Zone

Histochemical and electron microscopic evidence demonstrates the presence on the external side of the membrane of a cell coat or "fuzz" which consists of elongate glycoprotein and glycolipid molecules extending toward, and probably interdigitating with, similar molecules extending from the membrane of the adjacent cell (see Fig. 2). The intercellular space, some 200 Å wide, far from being a simple saline milieu, contains a variety of macromolecules, bonded more or less strongly to cell membranes, and other polysaccharide molecules not so bonded, corresponding to the ground substance of extracellular space in connective tissue (but lacking collagen). Fixed negative charges, especially of sialic acid, abound in the conjugated carbohydrate material of the outer zone and may be important in the interaction of multivalent cations, particularly Ca^{++}, with the cell membrane and with macromolecules of intercellular space.

Molecular Recognition: Specificity of Intercellular Adhesion

In the orderly development of an organism, especially of its most complex organ, the brain, it must be assumed that, under precise genetic control, macromolecular "tags" are synthesized, find their way to the limiting envelope of the cells, and there promote binding of similar or complementary molecules on the surface of adjacent cells whose biological role is to function in close concert.

One of the most graphic illustrations of such specific interactions is that posed by the experiment of SPERRY [48] in which, after cutting the optic nerve, fibers from the retina were shown to be able to join with their appropriate fibers of the optic nerve and to do so in such a manner that the functional tectal connectivity was again restored. This type of experimentation has recently been explored in more detail by GAZE and JACOBSON [49] (see also GAZE [50]), and a gradient hypothesis has been offered by JACOBSON [51, 52] to explain the dimensional and temporal specificity.

A striking example of specificity of cell-cell interaction temporally and spatially is that provided by the beautiful recent experiments of R. L. SIDMAN and his collaborators at the Harvard Medical School [53, 54]. Because of the orthogonal, "printed-circuit" synaptic connectivity of the major cell types of the cerebellar cortex, so organized as to permit the cerebellum to control and regulate every detail of muscular coordination of bodily activity, the cerebellum provides an especially favorable nervous tissue in which to study factors that determine membrane specificity in neurogenesis.

On the 18th prenatal day the mouse cerebellum and cerebral cortex are already well organized. If at this time (not at 17 or 19 days!) the cerebellar cells are dissociated mechanically in the presence of trypsin and allowed to round up by the use of a swirling flask technique, it is possible, after stopping the tryptic action, to arrange conditions so that the various types of cells migrate in orthogonal array with spatial relationships characteristic of the normal cerebellum — a truly remarkable example of specification of three-dimensional cellular movement and specific cell-cell surface recognition and adhesion.

Among the several types of neurally mutant mice that have been characterized is the "reeler", so called because of the incoordination of muscular movements (ataxia) due to failures in development of the cerebellum. The reeler's prenatal cerebellum can be dissociated with trypsin and, after inactivation of the trypsin, reconstituted; the result of the reconstitution is not a normal cerebellar connectivity, but that characteristic of the reeler mutant. Since the mutation is thought to be a point mutation, one protein — perhaps an enzyme — may be responsible for the abnormal connection of neuron types. This type of *in vitro* experimentation of specification of complex neural interconnection permits biochemical investigations of molecular mechanisms involved. These phenomena provide an exciting challenge to molecular membranologists and offer excellent opportunities for the discovery of membrane surface structures, their genetic, epigenetic, and developmental specification.

Specificity of macromolecular interaction or "recognition" is thought by some to underlie plasticity of synaptic connectivity basic to learning and memory. On the basis of analytical evidence, BOGOCH [55—57] suggests that polysaccharides, glycoproteins, and glycolipids are the substrates of this activity and that the transferase enzymes that mediate the addition of monosaccharide units play a central role in the process.

Experimental evidence adduced by BARONDES [58, 59] demonstrates that, in nerve terminals, carbohydrate residues are added to polypeptides synthesized in the neuronal cell body and transported to the terminals. Glycoproteins are thus synthesized during specific periods in CNS development. He has also proposed a gradient model for interneuronal recognition based on complementarity of cell surfaces in the retino-tectal system.

Relatively little is known concerning the biosynthesis and the deployment of glycoproteins and glycolipids at the cell surface membrane, although detailed studies have been made with microorganisms and with intracellular processes (see HAKOMORI and MURAKAMI [60], and HAKOMORI [61]). DORFMAN [62, 63] has contributed much to this subject over the years. Among other striking points that he mentions is the necessity for a template or cluster of monosaccharide transferases, in which the sequence of enzymes reflects the sequence of saccharides, to be added to the oligosaccharide chains in glycoproteins. Fig. 9 shows such a cluster of seven transferases. It is believed that such addition of oligosaccharides occurs as a terminal process in the ribosomal synthesis of polypeptide. The idea that glycosyl transferases are present in substantial concentrations at axon terminals has been suggested by ROSEMAN [64] and collaborators [65, 66]. They believe that terminal β-D-galactopyranosyl groups at ends of oligosaccharide chains of glycoproteins and mucins are present in the outer cell coat and are involved in specific adhesion of cells. The experimental basis of these claims is the estimation of states of aggregation by application of a Coulter counter and by ^{32}P labeling of suspensions of isolated neural retinal cells.

ROSEMAN [64] has discussed various theories of specific adhesion of molecules at the surfaces of adjacent cells. These include: 1) the antigen-antibody model of TYLER [67] and WEISS [68]; 2) the formation of H-bonds between saccharide units of oligosaccharide chains protruding from each cell and forming a double-chain complex whose tertiary conformation is unknown (conceivably it might follow a logic similar to the H-bonding of nucleotide residues in the double helix of DNA); and 3) the binding of a transferase enzyme on one cell by an oligosaccharide chain from another cell. ROSEMAN favors the third-mentioned theory, but points out that, in the presence of an appropriate monosaccharide and UDP as coenzyme, the transferase will become functional, thus breaking the enzyme-substrate bond that is responsible for the specific intercellular adhesion. This speculative notion is mentioned here because of the growing evidence for the presence of transferase enzymes in the surface membranes of cells.

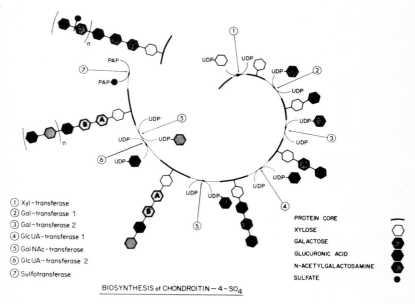

Fig. 9. Pathway of biosynthesis of proteochondroitin sulfate. From DORF-MAN [63]

Specific Ion Permeability and Bioelectric Properties of Membranes

The notion that solutes enter cells passively through non-specific pores or channels, hydrophilic or lipophilic, is being replaced by the concept that solutes are conveyed across the membrane bound to permease ligands more or less specific for the solute in question. Much advance has been made in the characterization of transport systems for non-electrolytes such as sugar, but our concern here will be for the case of carriers for electrolytes (ionophores) particularly involved in bioelectric processes.

According to the ionic theory of nerve action, the inward permeability for Na^+ is increased in the rising phase of the action potential and that for K^+ is increased during the falling phase of the action potential, the entire process being over in 10^{-4} seconds in a fast axon. The question "How are Na^+ and K^+ distinguished by cells?" is vital for an understanding not only of excitable membranes, but of cell membranes and organelles generally. It appears that a solution may be in sight for this classical biological problem.

Models of the molecular mechanism of the nerve action potential are of two main classes:

1) That which postulates a fast, two stable state conformational change of macromolecules arrayed in monolayers within the membrane (TASAKI [69], CHANGEUX et al. [70], BLUMENTHAL et al. [71], PODLESKI and CHANGEUX [72]); there is little direct evidence supporting this view, although ADAM [73] has put forward calculations consistent with this theory.

2) That which suggests that ions travel through ion-specific channels (HODGKIN [74], see also COLE [75]) that are sparsely (ca. $10-20/\mu^2$) distributed in the membrane (see KEYNES [76]).

It was hoped that fast optical methods, such as changes of birefringence, light scattering, or fluorescent probes, might be used to detect macromolecular conformational changes (SCHMITT and SCHMITT [77], KEYNES [76], COHEN et al. [78]). Changes were indeed observed, but these proved voltage-dependent, rather than current-dependent, and therefore probably relate to changes other than specific cation gating (COHEN and KEYNES [79]). KEYNES suggested that the number of channels may be too few and the optical effects too weak to be detected by methods thus far developed. Most recently, COHEN et al. [80] localized the change in retardation, which has a time course similar to that of the action potential, to a thin cylinder immediately surrounding the axoplasm of the squid giant axon and attributed it to material manifesting a radial optic axis. The nature of the hypothesized material is unknown, but is thought to consist of rather small molecules.

Lipid bilayer models have been developed that have properties of ion transport similar to those observed in natural membranes.

In their early, trail-blazing work, MUELLER and RUDIN [81] found that the high resistance of phospholipid bilayers could be reduced to physiological levels by addition to the solution surrounding the

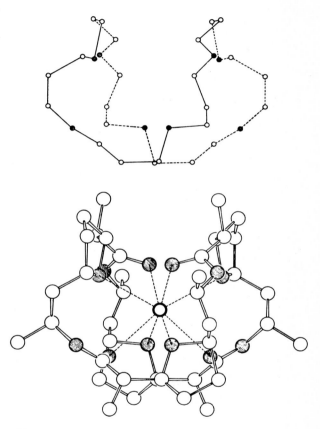

Fig. 10. The nonactin-K$^+$ complex viewed down the b crystal axis. From KILBOURN et al. [85]

bilayer of an "excitation imparting material" (EIM) of unspecified nature. Later these investigators [82] showed that the antibiotic alamethesin and certain peptides confer on the film the properties of rectification and the ability to display "action currents" and rhythmic discharges.

Many ionophores have now been studied (see particularly TOSTESON [83], EIGEN and DE MAEYER [84]), including the cyclic depsipeptides (i.e., peptides containing α-aminoacids and oxyamino-

IONOPHORE MECHANISM

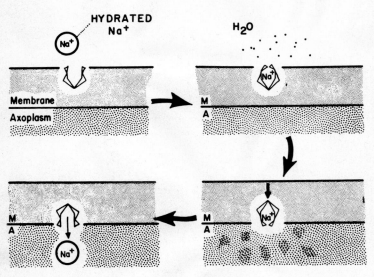

Fig. 11. Diagrammatic representation of peptide ion carrier to indicate role of change in solvation of ion and in conformation of peptide carrier. Compare with Fig. 10

acids), macrotetrolides, and certain so-called "Crown" compounds. Crystallographic studies (KILBOURN et al. [85]) have disclosed a structural basis for the entrapment of metal cations into the peptide cage. An example is shown in Fig. 10 for nonactin. Fig. 11 indicates diagrammatically the process described by EIGEN and WINKLER [86] and EIGEN and DE MAEYER [84] by which the cation is stripped of its solvate water molecules seriatim, one by one, as the ion enters the peptide carrier cage and bonds coordinately with oxygen atoms of the cage. Closure of the cage occurs by conformational change. Transport of the peptide-enclosed ion through the lipid phase of the membrane is facilitated by the fact that the externally directed side-chains of the peptide are predominantly hydrophobic. Binding of cations may vary among peptide ionophores by several orders of magnitude, suggesting a mechanism for specificity of ion transport and a possible answer to a central problem of molecular biology, i.e., how cells distinguish Na^+ and K^+ which have such similar physical properties.

EIGEN and WINKLER [86], working with murexide and the nonactins, have listed properties that characterize an efficient specific metal ion carrier: possession of electrophilic groups that can compete with solvent molecules for metal ion binding; close adaptation of the carrier cavity to the size of the metal ion; optimal fit leading to maximization of free energy of ligand binding and solvation; conformational flexibility sufficient to permit *stepwise* binding of metal ion and substitution of solvate molecules; and fast (ca. 10^{-8} sec.) loading and unloading of the metal ion, making such ionophores candidates for ion gating of Na^+ and K^+ in fast bioelectric processes such as the propagation of nerve action-potentials which may have time constants of the order of 10^{-4} seconds. E. GRELL (personal communication), in EIGEN's laboratory, using sound absorption and temperature-jump methods, has found, for the cyclodepsipeptides, valinomycin and enniatin B, two or three relaxation times having constants of the order of 10^{-7} to 10^{-9} seconds, corresponding to stepwise conformational changes in the opening of the cage; these rate-limiting conformational changes are characteristically different for Na^+ and K^+.

The ionophore mechanism contrasts with the ion-specific "channell" concept of HODGKIN, HUXLEY, and KEYNES for which no detailed molecular and physicochemical specifications have yet been proposed. It is not impossible that membrane proteins might contain portions or subunits which function in a manner comparable to that of ionophores carrying their ionic cargo through the membrane. Ion-specific pores that are not carriers have been described by G. EISENMAN et al. [87] and by D. W. URRY [23].

Peptide ionophores have not yet been demonstrated to exist in cell membranes, and it may be difficult to prove their presence because their composition is similar to proteinaceous material generally and because the ionophores may be present in such low concentrations in membranes as to escape detection by the most sensitive optical methods thus far devised.

From the sparsity of ion channels in the axonal membrane (ca. $10\text{---}20/\mu^2$), according to Hodgkin and Huxley [88], one may get the notion that, if we could look down at the cell membrane with molecular spectacles, we might see chiefly open territory, represented by the lipid-protein matrix of the membrane; molecular machines, such as receptors, carriers or channels, pumps, recognition molecules, occur relatively sparsely. Quite a different view is suggested by the recent work of Miledi et al. [89] who estimate that there are 10^{10} acetylcholine receptor binding sites per cell or ca. 10^4 sites per square micron of postsynaptic membrane. Sjöstrand and Barajas [90], from their electron microscopic studies, visualize the cell membrane as containing predominantly globular molecules interlarded with occasional lipid-bilayer regions.

An important question is whether there is in the cell membrane two- or three-dimensional interrelationship of molecules which have information processing capability.

Lehninger [91, 92], for example, finds as many as 20 to 30 polypeptides, many of which may be enzymes in the inner membranes of mitochondria. He is examining the possibility that microcybernetic regulatory properties may characterize clusters and networks of enzymes and other membrane-borne macromolecules. E. A. Boyse [93] visualizes elaborate molecular codes with regional distribution of surface antigens that can move about in the surface. The molecular surface display on membranes may be unique not only for each cell type, but perhaps for each individual cell. Determinants of these structures may be epigenetic rather than genetic, requiring relatively few genes to generate the constituent macromolecules which then self-assemble on the membrane in a fashion suggested by the experiments described by Rothfield and Pearlman-Kothencz [94] and by Rothfield [95].

Transduction and Amplification of Chemical Signals by Membrane-Borne Molecular Devices

Hormones, like other highly active biodynamic substances, such as synaptic transmitters, act in concentrations of the order of 10^{-8}

to 10^{-12} M. Present in these low concentrations in the blood, they present a "to-whom-it-may-concern" signal, i.e., only those cells having receptors for the particular hormone or transmitter respond. The puzzle, how the response to such signals could be orders of magnitude greater chemically than the signal, may now be nearer to solution through the discovery that an enzyme, adenyl cyclase, seems to be coupled with the hormone receptor, so that when the hormone binds to the receptor, the latter, by cooperative interaction, excites the adenyl cyclase to convert ATP in the cell to cyclic AMP. Cyclic AMP has a positive ΔF in equilibrium with ATP, and the enthalpy of hydrolysis of the 3′ bond of cyclic AMP is -14.1 Kcal/mole (GREENGARD et al. [96]). It acts as an intracellular molecular effector, adenylating protein, activating enzymes, through phosphorylation of kinase enzymes in many cases, and effecting glycogenolysis, lipolysis, steroidogenesis, enzyme induction, polypeptide secretion, contractile force, Na^+ extrusion, and other membrane permeability changes — a truly ubiquitous and global type of biological activity (see SUTHERLAND et al. [97] and, for the role of cyclic AMP in the nervous system, see RALL and GILMAN [98]). The action of cyclic AMP as a "second messenger" is illustrated graphically in Fig. 12.

Cells produce cyclic AMP in concentration in the range of 10^{-3} M. Since the hormone which stimulates its production may be in the range of 10^{-8} to 10^{-12} M, the amplification factor is 10^5 to 10^9!

The recent work of RODBELL et al. ([99—102], POHL et al. [103], and BIRNBAUMER et al. [104]) represents a significant step in understanding the transduction and amplification of chemical signals from hormones. RODBELL suggests that the receptor or *discriminator* occurs on the extracellular side of a solid state device situated in the membrane. Facing the intracellular side is the *amplifier* portion, i.e., the adenyl cyclase. Coupling the two in cooperative interaction is the *transducer* component, not yet chemically identified, but possibly a lipid protein component of the membrane. Fat cells can discriminate as many as nine hormones (glucagon, insulin, ACTH, TSH, oxytocin, vasopressin, growth hormones, corticosteroids, and thyroxin). These can all stimulate adenyl cyclase through a common type of transducer molecules. The transduction and amplification process is entrained only at the moment when the hormone binds to the receptor; the process is not a continuing one, but a one-shot, "quantum" reaction. ROD-BELL suggests that GTP is bound to a site on the receptor different from that which binds the hormone; GTP binding, by cooperative interaction, releases the hormone from its receptor binding-site. The mechanism is then "cocked" and ready to be activated again

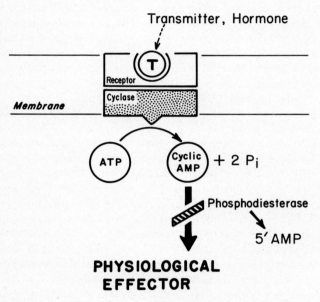

Fig. 12. Suggested solid-state, membrane-borne device (after RODBELL et al. [99]), consisting of receptor, membrane transducer, and adenyl cyclase, whereby hormone binding leads to production of cyclic AMP with corresponding great amplification of physiological effect

by binding of another hormone molecule, to start the whole receptor-transducer-cyclase excitation cycle over again. Such a mechanism would explain the turnover of hormones, which is known to be high.

RODBELL's speculation is an attractive one. Unfortunately, none of the three components of the solid state device has yet been purified and characterized chemically. For recent advances in knowledge of receptor properties, see PORTER and O'CONNOR [105]. Notions of the structure of at least one receptor, that for acetylcholine, have been derived from experiments of affinity labeling (see KARLIN [106, 107], DEL CASTILLO et al. [108, 109]). Direct isolation and chemical characterization of this receptor have been reported by MILEDI et al. [89], using α-bungarotoxin (from a Formosan snake), previously shown by CHANGEUX et al. [110] to bind strongly to acetylcholine receptors. The data are best interpreted on the assumption that the receptor protein is a tetramer; the molecular weight of the monomer containing one binding site turns out to be 80,000. Other suggestions about receptor structures have been offered by DE ROBERTIS [111].

Membrane Behavior in Cell Transformation and Malignancy

Exciting advances have been made recently in our understanding of the mechanism by which cells are transformed by viruses and carcinogenic hydrocarbons, resulting in changes in the cell's "social behavior" which triggers unbridled cell division and malignancy. These advances may lead to new concepts basic to the problem of cancer. The changes that occur in cell transformation include, according to STOKER [112], the following: loss of contact inhibition of movement and growth; loss of substrate anchorage dependence for growth; and changes in solute uptake and transport. With transformation, new membrane-bound antigenic determinants appear, at least one of which has been shown to be determined by the host genome (see particularly the interesting work of DULBECCO [113] and ECKHART et al. [114]) on activation by polyoma virus.

Viral transformation changes the amount and kinds of glycolipids and glycoproteins [60]. ROBBINS and MACPHERSON [115, 116] found several large glycolipids to be lacking in the surface membrane of transformed hamster cells, and this has been related to loss of contact inhibition. The mechanism of control of animal cell surfaces is yet little explored; most of what is known derived from studies of micro-organisms. ROBBINS and MACPHERSON suggest that glycolipids vary in composition among different tissues even more than do phospholipids and are closely controlled by genetic, physiologic, and hormonal influences. The extent to which transformation involves a particular subset of these glycolipid and glycoprotein variants remains to be determined.

Higher than normal agglutinating ability is conferred on cells by certain conjugated carbohydrates, called lectins, such as those extracted from the jack bean and wheat germ. One of these, concanavalin A, which has played an important role in studies of cell transformation, reacts through binding sites that contain lectin-specific sugars, particularly N-acetyl glucosamine. These sites normally become available only briefly during mitosis, but remain accessible in the membranes of transformed cells. Further experimentation and interpretation of concanavalin A effects will be facilitated by the work of G. M. EDELMAN and his colleagues [117], which includes analysis of composition and sequence of amino acids and X-ray diffraction analyses of the tertiary structure of the protein. From studies with circular dichroism, EDELMAN finds that conformational changes occur when sugars react with concanavalin A (personal communication, 1971).

Especially significant for this subject is the work of M. M. BURGER of Princeton University and L. SACHS at the Weizmann Institute, which relates changes in cell behavior with membrane changes observed after experimental blocking with concanavalin A of membrane sites that determine contact inhibition of cell division or by activation of sites, normally cryptic, by treatment with proteases such as very dilute trypsin. BURGER [118, 119] and BURGER and NOONAN [120] suggest that trypsin breaks few peptide bonds, only enough to rearrange the molecular architecture of the surface membrane sufficiently to unmask and activate the membrane sites that determine contact inhibition. Transformation of cells, leading to "malignancy" by treatment with oncogenic viruses or carcinogenic hydrocarbons may liberate intracellular proteases that unmask agglutinin-combining sites in the outer surface of the membrane. Treatment of such transformed cells with concanavalin A masks the surface-exposed sites and reestablishes contact inhibition of mitosis.

MOSCONA [121] has found that concanavalin A agglutinates embryonic cells which had been dissociated by EDTA. Availability of certain membrane sites is apparently required for differentiation and embryogenesis. One wonders whether the neuron may be a transformed neuroblast and whether specific synaptic agglutination is importantly determined by the properties of agglutinin-like sites and the presence of certain natural lectins or of proteases liberated perhaps by lysosomes, whose role in brain function deserves careful study.

Membrane-Gene Linkage

Examples from many fields of biology could be cited which involve linkage between the genetic mechanisms of the cell center and constituents in the cell's surface membrane. Space permits reference to only two systems: first, that controlling cell division, clearly related to the problems of contact inhibition of mitosis just discussed; and, secondly, that concerning the regulation of interneuronal interaction.

Fox et al. [122] relate control of mitosis to genetic events and propose that positive feedback loops govern cyclic function of the cell. These investigators stress the importance for normal cell function of maintenance and control of the molecular apparatus that subserves these membrane-nuclear linkages and feedback loops.

Several lines of current research demonstrate that synaptic activation of neurons results in an adaptive turning on of genetically controlled syntheses. PETERSON [123], PETERSON and KERNELL [124], and KERNELL and PETERSON [125], working with isolated abdominal ganglia of *Aplysia* incubated in a tissue bath, found that synaptic stimulation increased the rate of synthesis of RNA as measured by the incorporation of tritiated uridine and cytidine. This increase was not correlated with spike potentials elicited by antidromic invasion of the cell, hence to all-or-none depolarization of the neuronal membrane, but to events triggered at the postsynaptic membrane.

The stimulation in RNA production might be expected to increase synthesis of specific proteins. Although such results have not yet been obtained from *Aplysia*, AXELROD's group [126, 127] succeeded in the case of the rat to show that stimulation of adrenergic neurons induced in the postsynaptic cell increased rate of synthesis of tyrosine hydroxylase, the rate-limiting enzyme for the production of norepinephrine, the transmitter of adrenergic neurons. After synthesis in the neuronal cell center, the enzyme is translocated down the axon to the terminals where the major synthesis of norepinephrine occurs.

McIlwain [128] has pointed out that neurons contain a multiplicity of enzymes that can be induced through gene expression (at least in part as a response to membrane changes), and this large repertoire of enzymes with specific adaptive inducibility may be as important for brain processes, including ontogenetic specification and physiological and psychological plasticity, as is the brain's ability to form a multiplicity of synaptic interconnections.

Not only do postsynaptic processes lead to changes in gene expression, but also gene products, acting on the excitable membrane in the trigger zone (axon hillock) of the neuron, may modulate excitability adaptively and participate in the learning process.

An important research target is the discovery of the molecular mechanism for the two-way linkage or feedback loops between membrane-borne systems and the genetic apparatus, i.e., does the linkage occur alone by diffusion of molecular effectors such as cyclic AMP; or do specialized organelles, such as microtubules, convey materials rapidly in a spatially directed manner?

Concluding Remarks

The plasma membrane of many cell types contains a lipid-protein "floor space" that includes a substantial, although undetermined, fraction in the form of lipid bilayers, bounded, interpenetrated, and traversed by protein and other non-lipid constituents. In addition to this inner portion of the "greater" membrane, there is an outer portion containing glycoprotein, glycolipid, and other conjugated and unconjugated carbohydrate moieties essential for certain vital processes such as cell-cell recognition, contact inhibition of mitosis, and, in the brain, possibly for the determination of neuronal circuitry, plasticity, and learning.

The linkage between membrane and genetic control centers in the nucleus forms a feedback loop vital to functioning, but still poorly understood.

Mounted upon this intrinsic or "inner" membrane are molecular assemblies or "machines" such as receptors, transducers, amplifiers, permeases, and pumps. Some of these, e.g., the cyclic AMP and the concanavalin agglutinating systems, have been discovered only in the last decade; other important ones which doubtless exist remain to be discovered. Among these may be cybernetic topochemical informational macromolecular computer systems for homeostatic and adaptive intercellular communication, and, in the brain, for storage and retrieval of psychological information. A truly challenging prospect for biochemical and biophysical membranology!

References

1. PARDEE, A. B.: Abstract, pp. 9—10. In: Symposium on Membranes and the Coordination of Cellular Activities, sponsored by Biology Division, Oak Ridge National Laboratory, Gatlinburg, Tenn., April 5—8, 1971.
2. KATCHALSKY, A., OPLATKA, A.: In: Neurosciences Research Symposium Summaries, Vol. 1, 353—373, (SCHMITT, F. O., MELNECHUK, T., Eds.) Cambridge, Mass.: M. I. T. Press 1966.
3. — In: Membrane models and their formation of biological membranes, pp. 318—332. (BOLIS, L., PETHICA, B. A., Eds.) Amsterdam: North-Holland Publishing Co., and New York: John Wiley & Sons, Inc., Wiley Interscience Division 1968.
4. — OSTER, G.: In: The molecular basis of membrane function, pp. 1—44. (TOSTESON, D. C., Ed.). (Symposium of Society of General Physiologists, Durham, North Carolina, August 20—23, 1968.) Englewood Cliffs, N. J.: Prentice-Hall, Inc. 1969.
5. DELBRÜCK, M.: In: The Neurosciences: Second study program, pp. 677 to 684. (SCHMITT, F. O., Ed.) New York: Rockefeller University Press 1970.
6. REVEL, J.-P., ITO, S.: In: The specificity of cell surfaces, pp. 211—234. (DAVIS, B. D., WARREN, L., Eds.) Englewood Cliffs, N. J.: Prentice-Hall, Inc. 1967.
7. SCHMITT, F. O., SAMSON, F. E.: Brain cell microenvironment. Neurosciences research program bulletin 7 (4), 278—417 (1969); also: Neurosciences Research Symposium Summaries, 4, 191—325. (SCHMITT, F. O., MELNECHUK, T., QUARTON, G. C., ADELMAN, G., Eds.) Cambridge, Mass.: M. I. T. Press 1970.
8. ITO, S.: J. Cell Biol. 27, 475—491 (1965).
9. RAMBOURG, A., NEUTRA, M., LEBLOND, C. P.: Anat. Rec. 154, 41—71 (1966).
10. — J. Histochem. Cytochem. 15, 409—412 (1967).
11. ITO, S.: Fed. Proc. 28, 12—25 (1969).
12. DANIELLI, J. F., DAVSON, H.: J. cell. comp. Physiol. 5, 495—508 (1935).
13. BEAR, R. S., SCHMITT, F. O.: J. Opt. Soc. Amer. 26, 206—212 (1936).
14. SCHMIDT, W. J.: Die Doppelbrechung von Karyoplasma, Zytoplasma und Metaplasma. Berlin: Verlag von Gebrüder Borntraeger 1937.
15. SCHMITT, F. O., BEAR, R. S., CLARK, G. L.: Radiology 25, 131—151 (1935).
16. — — PALMER, K. J.: J. cell. comp. Physiol. 18, 31—42 (1941).
17. BEAR, R. S.: In: Myelin. Neurosciences research program bulletin 9 (4), 507—570 (1971). (MOKRASCH, L. C., BEAR, R. S., SCHMITT, F. O., Eds.).

18. DANIELLI, J. F., GREEN, D. E. Chairmen: Conference on Membrane Structure and its Biological Applications. New York, June 2—4, 1971. New York: New York Akademy of Sciences (in press).
19. MCCONNELL, H. M.: In: Neurosciences: Second study program, pp. 697 to 706. (SCHMITT, F. O., Ed.) New York: Rockefeller University Press 1970.
20. WALLACH, D. F. H., ZAHLER, P. H.: Proc. nat. Acad. Sci. (Wash.) **56**, 1552—1559 (1966).
21. — GORDON, A.: Fed. Proc. **27** (6), 1263—1268 (1968).
22. SINGER, S. J.: In: Membrane structure and function (in press). (ROTHFIELD, L. I., Ed.) New York: Academic Press 1971.
23. URRY, D. W.: In: Conference on Membrane Structure and its Biological Applications. Chairmen: DANIELLI, J. F., GREEN, D. E. New York, June 2—4, 1971. New York: New York Academy of Sciences (in press).
24. ROTHFIELD, L., FINKELSTEIN, A.: Ann. Rev. Biochem. **37**, 463—496 (1968).
25. METCALFE, J. C.: Abstracts (This Symposium).
26. FISCHER, S., CELLINO, M., ZAMBRANO, F., ZAMPIGHI, G., NAGEL, M. T., MARCUS, D., CANESSA-FISCHER, M.: Arch. Biochem. Biophys. **138**, 1—15 (1970).
27. HUBBELL, W. L., MCCONNELL, H. M.: J. Amer. Chem. Soc. **93**, 314—326 (1971).
28. MCFARLAND, B. G., MCCONNELL, H. M.: Proc. nat. Acad. Sci. (Wash.) **68**, 1274—1278 (1971).
29. KORNBERG, R. D., MCCONNELL, H. M.: Biochemistry **10**, 1111—1120 (1971).
30. TRÄUBLE, H.: J. Membrane Biol. **4**, 193—208 (1971).
31. CHAPMAN, D.: The structure of lipids. London: Methuen 1965.
32. — WALLACH, D. F. H.: In: Biological membranes, pp. 125—202. (CHAPMAN, D., Ed.). London and New York: Academic Press 1968.
33. PHILLIPS, M. C., WILLIAMS, R. M., CHAPMAN, D.: Chem. Phys. Lipids **3**, 234—244 (1969).
34. LADBROOKE, B. D., CHAPMAN, D.: Chem. Phys. Lipids **3**, 304—356 (1969).
35. LEHNINGER, A. L.: In: Neurosciences Research Symposium Summaries, Vol. **1**, 294—317 (SCHMITT, F. O., MELNECHUK, T., Eds.) Cambridge, Mass.: M. I. T. Press 1966.
36. GREEN, B. B.: Exp. Cell Res. **7**, 558—562 (1954).
37. AGRAWAL, H. C., BANIK, N. L., BONE, A. H., DAVISON, A. N., MITCHELL, R. F., SPOHN, M.: Biochem. J. **120**, 635—642 (1970).
38. O'BRIEN, J. S.: Science **147**, 1099—1107 (1965).
39. — SAMPSON, E. L.: J. Lipid Res. **6**, 545—551 (1965).
40. — J. theor. Biol. **15**, 307—324 (1967).
41. FINEAN, J. B., MILLINGTON, P. F.: J. biophys. biochem. Cytol. **3**, 89—94 (1957).
42. ROBERTSON, J. D.: J. biophys. biochem. Cytol. **4**, 349—364 (1958).
43. — In: Cellular membranes in development, pp. 1—81 (LOCKE, M., Ed.). New York: Academic Press 1964.

44. BLAUROCK, A. E., WORTHINGTON, C. R.: Biochim. biophys. Acta (Amst.) **173**, 419—426 (1969).
45. CASPAR, D. L. D., KIRSCHNER, D. A.: Nature New Biology **231**, 46—52 (1971).
46. BRANTON, D.: Exp. Cell Res. **45**, 703—707 (1967).
47. PARK, R. B.: In: Conference on Membrane Structure and its Biological Applications. Chairmen: DANIELLI, J. F., GREEN, D. E. New York, June 2—4, 1971. New York: New York Academy of Sciences (in press).
48. SPERRY, R. W.: J. Neurophysiol. **7**, 57—70 (1970).
49. GAZE, R. M., JACOBSON, M.: Proc. roy. Soc. B **157**, 420—448 (1963).
50. — The formation of nerve connections, 288 pp. London, New York: Academic Press 1970.
51. JACOBSON, M.: Developmental neurobiology, 465 pp. New York: Holt, Rinehart and Winston, Inc. 1970.
52. — In: The neurosciences: Second study program, pp. 116—129 (SCHMITT, F. O., Ed.). New York: Rockefeller University Press 1970.
53. SIDMAN, R. L.: In: The neurosciences: Second study program, pp. 100 to 107 (SCHMITT, F. O., Ed.) New York: Rockefeller University Press 1970.
54. DELONG, G. R., SIDMAN, R. L.: Develop. Biol. **22**, 584—600 (1970).
55. BOGOCH, S.: In: Brain and nerve proteins: Functional correlates. Neurosciences research program bulletin **3** (6), 38—41 (1965). (SCHMITT, F. O., DAVISON, P. F., Eds.). See also: Neurosciences Research Symposium Summaries, **2**, 374—376 (1967). (SCHMITT, F. O., MELNECHUK, T., QUARTON, G. C., ADELMAN, G., Eds.). Cambridge, Mass.: M. I. T. Press 1967.
56. — The biochemistry of memory. New York: Oxford University Press 1968.
57. — Neurosciences research program bulletin **7** (4), 351—354 (1969). See also: Neurosciences Research Symposium Summaries **4**, 265—268 (1970).
58. BARONDES, S. H.: J. Neurochem. **15**, 699—706 (1968).
59. — In: The neurosciences: Second study program, pp. 747—767 (SCHMITT, F. O., Ed.). New York: Rockefeller University Press 1970.
60. HAKOMORI, S.-I., MURAKAMI, W. T.: Proc. nat. Acad. Sci. (Wash.) **59**, 254—261 (1968).
61. — Abstracts (This Symposium).
62. DORFMAN, A.: In: Connective tissue: Intercellular macromolecules, pp. 155—165 (Proceedings of Symposium sponsored by the New York Heart Association). Boston: Little Brown and Co. 1964.
63. — In: Chemistry and molecular biology of the intercellular matrix, Vol. **3**, pp. 1421—1448 (BALAZS, E. A., Ed.). London and New York: Academic Press 1970.
64. ROSEMAN, S.: Chem. Phys. Lipids **5**, 270—297 (1970).
65. ROTH, S., MCGUIRE, E. J., ROSEMAN, S.: Intercellular adhesive specificity. J. Cell Biol. (November 1971) (in press).
66. — — — Evidence for cell-surface glycosyltransferases: their potential role in cellular recognition. J. Cell Biol. (November 1971) (in press).

67. TYLER, A.: Growth (Suppl.) **10**, 7—19 (1946).
68. WEISS, P.: Yale J. Biol. Med. **19**, 235—278 (1947).
69. TASAKI, I.: Nerve excitation: a macromolecular approach. Springfield, Ill.: C. C Thomas 1968.
70. CHANGEUX, J.-P., THIERY, J., TUNG, Y., KITTEL, C.: Proc. nat. Acad. Sci. (Wash.) **57**, 335—341 (1967).
71. BLUMENTHAL, R., CHANGEUX, J.-P., LEFEVER, R.: J. Membrane Biol. **2**, 351—374 (1970).
72. PODLÉSKI, T. R., CHANGEUX, J.-P.: In: Fundamental concepts in drug-receptor interactions, pp. 93—119. (DANIELLI, J. F., MORAN, J. F., TRIGGLE, D. J., Eds.). New York and London: Academic Press 1970.
73. ADAM, G.: In: Physical principles of biological membranes, pp. 35—44. (SNELL, F., WOLKEN, J., IVERSON, G. J., LAM, J., Eds.). (Proceedings of Coral Gables Conference, Miami, December 18—20, 1968.) New York: Gordon & Breach Science Publishers 1970.
74. HODGKIN, A. L.: Science **145**, 1148—1153 (1964).
75. COLE, K. S.: Membranes, ions and impulses. Berkeley and Los Angeles: University of California Press 1968.
76. KEYNES, R. D.: In: The neurosciences: Second study program, pp. 707 to 714 (SCHMITT, F. O., Ed.). New York: Rockefeller University Press 1970.
77. SCHMITT, F. O., SCHMITT, O. H.: J. Physiol. (Lond.) **98**, 26—46 (1940).
78. COHEN, L. B., KEYNES, R. D., HILLE, B.: Nature (Lond.) **218**, 438—441 (1968).
79. — — J. Physiol. (Lond.) **204**, 100—101 P (1969).
80. — HILLE, B., KEYNES, R. D.: J. Physiol. (Lond.) **211**, 495—515 (1970).
81. MUELLER, P., RUDIN, D. O.: J. theor. Biol. **4**, 268—280 (1963).
82. — — J. theor. Biol. **18**, 222—258 (1968).
83. TOSTESON, D. C. (Ed.): The molecular basis of membrane function. (Symposium of Society of General Physiologists, Durham, N. C., August 20—23, 1968.) Englewood Cliffs, N. J.: Prentice-Hall, Inc. 1969.
84. EIGEN, M., DE MAEYER, L.: Carriers and specificity in membranes. Neurosciences research program bulletin **9** (3), 300—347 (1971).
85. KILBOURN, B. T., DUNITZ, J. D., PIODA, L. A. R., SIMON, W.: J. molec. Biol. **30**, 559—563 (1967).
86. EIGEN, M., WINKLER, R.: In: The neurosciences: Second study program, pp. 685—696 (SCHMITT, F. O., Ed.). New York: Rockefeller University Press 1970.
87. EISENMAN, G., SANDBLOM, J. P., WALKER, J. L., Jr.: Science **155**, 965—974 (1967).
88. HODGKIN, A. L., HUXLEY, A. F.: J. Physiol. (Lond.) **117**, 500—544 (1952).
89. MILEDI. R., MOLINOFF, P., POTTER, L. T.: Nature (Lond.) **229**, 554—557 (1971).
90. SJÖSTRAND, F. S., BARAJAS, L.: J. Ultrastruct. Res. **32**, 293—306 (1970).

91. LEHNINGER, A. L.: Abstract, pp. 28—29. In: Symposium on Membranes and the Coordination of Cellular Activities, sponsored by Biology Division, Oak Ridge National Laboratory, Gatlinburg, Tenn., April 5—8, 1971.
92. — Abstracts (This Symposium).
93. BOYSE, E. A.: Abstract, pp. 42—43. In: Symposium on Membranes and the Coordination of Cellular Activities, sponsored by Biology Division, Oak Ridge National Laboratory, Gatlinburg, Tenn., April 5—8, 1971.
94. ROTHFIELD, L., PEARLMAN-KOTHENCZ, M.: J. molec. Biol. **44**, 477—492 (1969).
95. ROTHFIELD, L. I.: Abstracts (This Symposium).
96. GREENGARD, P., RUDOLPH, S. A., STURTEVANT, J. M.: J. biol. Chem. **244**, 4798—4800 (1969).
97. SUTHERLAND, E. W., ROBISON, G. A., BUTCHER, R. W.: Circulation **3**, 279—306 (1968).
98. RALL, T. W., GILMAN, A. G.: The role of cyclic AMP in the nervous system. Neurosciences research program Bulletin 8 (3), 221—323 (1970).
99. RODBELL, M., BIRNBAUMER, L., POHL, S. L., KRANS, H. M. J.: Acta diabet. lat. **7** (Suppl. 1), 9—57 (1970).
100. — KRANS, H. M.-J., POHL, S. L., BIRNBAUMER, L.: J. biol. Chem. **246**, 1861—1871 (1971).
101. — — — — J. biol. Chem. **246**, 1872—1876 (1971).
102. — BIRNBAUMER, L., POHL, S. L., KRANS, H. M.-J.: J. biol. Chem. **246**, 1877—1882 (1971).
103. POHL, S. L., BIRNBAUMER, L., RODBELL, M.: J. biol. Chem. **246**, 1849—1856 (1971).
104. BIRNBAUMER, L., POHL, S. L., RODBELL, M.: J. biol. Chem. **246**, 1857—1860 (1971).
105. PORTER, R., O'CONNOR, M. (Eds.): Molecular properties of drug receptors. (Ciba Foundation Symposium, January 27—29, 1970). London: J. & A. Churchill 1970.
106. KARLIN, A.: J. gen. Physiol. **54**, 245—264 (1969).
107. — In: Macromolecules in synaptic function. Neurosciences research program bulletin 8 (4), 390—395 (1970). (BLOOM, F. E., IVERSEN, L. L., SCHMITT, F. O., Eds.).
108. DEL CASTILLO, J., ESCOBAR, J., GIJÓN, E.: Int. J. Neurosci. **1**, 199—209 (1971).
109. — — — On the physiological significance of the sulfhydryl groups of proteins in synaptic transmission and other excitation processes. In: Festschrift honoring Dr. Chandler Brooks (in press).
110. CHANGEUX, J.-P., KASAI, M., CHEN-YUAN, L.: Proc. nat. Acad. Sci. (Wash.) **67**, 1241—1247 (1970).
111. DE ROBERTIS, E.: Science **171**, 963—971 (1971).
112. STOKER, M. G. P.: Abstract, pp. 46—48. In: Symposium on Membranes and the Coordination of Cellular Activities, sponsored by Biology Division, Oak Ridge National Laboratory, Gatlinburg, Tenn., April 5—8, 1971.

113. DULBECCO, R.: Nature (Lond.) **227**, 802—806 (1970).
114. ECKHART, W., DULBECCO, R., BURGER, M. M.: Proc. nat. Acad. Sci. (Wash.) **68**, 283—286 (1971).
115. ROBBINS, P. W., MACPHERSON, I. A.: Proc. roy. Soc. **B 177**, 49—58 (1971).
116. — — Nature (Lond.) **229**, 569—570 (1971).
117. WANG, J. L., CUNNINGHAM, B. A., EDELMAN, G. M.: Proc. nat. Acad. Sci. (Wash.) **68**, 1130—1134 (1971).
118. BURGER, M. M.: In: Permeability and function of biological membranes, pp. 107—119 (BOLIS, L., KATCHALSKY, A., KEYNES, R. D., LOEWENSTEIN, W. R., BETHICA, B. A., Eds.). Amsterdam: North-Holland Publishing Co. 1970.
119. — Nature (Lond.) **227**, 170—171 (1970).
120. — NOONAN, K. D.: Nature (Lond.) **228**, 512—515 (1970).
121. MOSCONA, A. A.: Science **171**, 905—907 (1971).
122. FOX, T. O., SHEPPARD, J. R., BURGER, M. M.: Proc. nat. Acad. Sci. (Wash.) **68**, 244—247 (1971).
123. PETERSON, R. P.: J. Neurochem. **17**, 325—338 (1970).
124. — KERNELL, D.: J. Neurochem. **17**, 1075—1085 (1970).
125. KERNELL, D., PETERSON, R. P.: J. Neurochem. **17**, 1087—1094 (1970).
126. MUELLER, R. A., THOENEN, H., AXELROD, J.: J. Pharmacol. Exp. Therap. **169**, 74—79 (1969).
127. THOENEN, H., MUELLER, R. A., AXELROD, J.: J. Pharmacol. Exp. Ther. **169**, 249—254 (1969).
128. MCILWAIN, H.: Nature (Lond.) **226**, 803—806 (1970).

Contacts and Communications Between Cells in Their Relationship to Morphogenesis and Differentiation

R. AUERBACH

Department of Zoology, University of Wisconsin, Madison, WI 53706/USA

With 3 Figures

Two phenomena central to embryological thinking have been the processes of regulation and of inductive tissue interactions. Both of these represent embryological communication between cells. In the case of regulation, the cells of a mass must let each other know where they are in relation to each other so that in some way the total group of cells differentiates as a harmonious single structure. In the instance of embryonic induction, one group of cells influences another group, usually of a dissimilar type, to differentiate in a new direction, and this induction usually is reciprocal in manner. BONNER has used the term "chemical conversation" to describe the communication system of differentiating cells [1], a term which seems ideal since it portrays a picture without defining the medium used to create it [2].

The entire ontogenetic history of an individual can be viewed as one long series of cellular communications beginning with the complex events surrounding interaction between sperm and egg cells, and ending with the long-range communication systems of the adult which must involve not only cell-cell interactions of a direct type but a whole series of humoral influences mediated through the circulation to act at long distances. Implied in this continued series of cellular communications throughout ontogeny is the fact that methods of communication must of necessity change as the embryo increases in size and complexity [2, 3].

When one examines the first of these developmental interactive systems, that of sperm-egg interaction, one finds that it involves one of the most elaborate and exquisitely sensitive sequences of events known to biologists [4, 5]. A full review is clearly beyond the scope of the present paper but a few of the most important aspects of that interaction should at least be mentioned. Sperm-egg fusion involves specific cell surface components with unique antigenic properties; activity of antibody-like materials with some type of specific complementary reaction between sperm-produced and egg-produced materials has been demonstrated to be involved in some species at least; fertilization results in the release of cortical granules, changes in cell permeability, and specific immunity to further activation or penetration; cell fusion leads to the introduction of genetic material, differential activation of specific synthetic pathways, unmasking of blocked genetic information, induction of cell division and initiation of cellular differentiation.

The striking similarity between this type of interaction and the interaction seen in immunological differentiation has prompted an extensive discussion of the analogous events in the cellular interactions seen during gametogenesis and fertilization, and the events that are known to take place during the initial differentiation and subsequent interactions of lymphoid cells participating in immunological reactions [6]. We will return to this point later.

More than 30 years ago, in a classical embryological study, it was pointed out [7] that cells of an embryo have specific affinities and can recognize themselves as like, other cells as different; cells can orient in typical, appropriate fashion after disruption. This concept of "Gewebsaffinität" has received increased attention in recent years with the establishment of techniques for obtaining critical cell dissociation and reaggregation, and the development of tissue culture techniques adequate to permit critical analysis of the resultant reorganization of cells [8]. Again, a full review here is not feasible. The important features of *cell aggregation systems* are that they are quite specific, that cells change in their aggregation patterns and preferences during ontogeny, that aggregation properties are characteristic and predictable, that aggregation involves both highly specific and more general adhesive functions, and that aggregation tendencies change with age, differentiation, viral infection and physiological state of the cell. Recent work involving specific exudates suggests that materials, probably glycoprotein in nature, are released by embryonic cells which can promote specific aggregation; moreover, antibodies direct against specific exudates appear to block selectively the normal aggregation processes of those cells that can respond to these exudates [9, 10].

That aggregation tendencies play a role in adult life has become apparent as methods for cell and tissue culture of immunologically active cells have become available [11, 12, 13]. Cell cultures obtained from adult spleens can be induced to antibody formation only under those conditions which permit cell aggregation to proceed, and it has been shown that the formation of these aggregates include specific sorting out and tissue reconstruction prerequisite to the induction of specific antibody formation [14, 15].

A form of cell aggregation, or at least a phenomenon closely related to it, is that of *cell migration*. Cell migration plays a major role in the construction of the embryo, permitting not only the establishment of the primary embryonic axis, but also the formation of the heart anlage, the movement of the thymus to the thoracic cavity, and the establishment of ganglia. The pigment cells of the skin migrate there from the neural crest, the germ cells migrate to the gonad from the germinal ridge, and the entire hematopoietic system involves a complex series of migrations that include the yolk sac, bone marrow, liver, and the various peripheral erythroid and lymphoid organs. Little is yet known about the communication that must be taking place between the cells that migrate, the terminal location of those cells, and the intermediate underlying cellular substratum; why, for example, lymphoid cells move to the spleen, erythroid precursors to the liver, or germinal cells to the gonad, or why retinal fibers attach in sequential and orderly fashion to the topic tectum is not at all understood, yet presents an aspect of cellular communication that must be fundamental in the spectrum of events we are considering. It should, moreover, be kept in mind that cell migrations, so prevalent during embryogeny, continue to play a major role in the functioning adult, as seen for example in the hematological system and the immunological system which will be discussed in more detail in a later portion of this paper.

Tissue interactions, or embryonic inductions, represent a logical means for the embryo to develop as an integrated harmonious whole. They occur almost ubiquitously throughout ontogeny, as each organ becomes shaped into its definitive form. Tissue interactions exist not only for the classical induction of the primary axis and nervous system, but also for the development of skin, hair, teeth, lung, thymus, pancreas, salivary gland, lens, thyroid, in short for every system which has been analyzed in depth [16, 17, 18].

Unfortunately, in spite of years of analysis, and in spite of the use of sophisticated techniques of isotopic labelling, electron microscopy, graded filter interposition, and extensive biochemical analysis, we still have no real clue concerning the nature of specific inductively active materials which mediate the specific differentiative events. Only some general characteristics can really be stated with confidence. Inductive interactions usually occur between dissimilar tissues, such as epithelium and mesenchyme, or epithelium, nervous tissue and mesenchyme. Operationally, there is a requirement for living cells, close proximity, with an intervening millipore barrier restricted to a maximum distance of 80 μ and a minimum pore size of 0.1 μ. Induction normally includes activation of new synthetic processes and the initiation of cell division. While for specific inductive interactions a "chemical inducer" may have been identified, such as "nucleic acid", "oligonucleotides", "carbon dioxide" or "small protein", as yet there is no discernible pattern, and the published identifications of inducers have been less than convincing even for the specific inductions under analysis [19].

Inductive interactions continue to play a significant role throughout adult life, both in the maintenance of the differentiated state, and in repair and regeneration. The operating rules for induction in adult systems appear similar to those seen in embryos in terms of specificity, initiation of new synthesis, induction of cell division, and cross-reactivity. An additional role is played by circulating humoral factors, such as, for example, erythropoietin, which can effect hematopoiesis of bone marrow cells, but fails to influence the early embryonic hematopoietic stem cells. Various other humoral factors of chalones have been shown to influence specific adult inductive systems, but again in a manner whose nature has not at all been delineated beyond the descriptive level.

With this background concerning cellular communication in developmental systems I would like to turn specifically to the *development of the immune system,* as defined to include both the ontogeny of the participating cells and the development of a specific immune response in reaction to the introduction of antigen. This latter area of cellular communication or cell interaction in immune systems has received much attention in recent years [20], but will be viewed here as merely a logical extension of and example of the type of interactive system shown to be involved in virtually all developmental systems.

Cell migrations in the immune system include the migrations characteristic of early development, the migrations that occur during normal adult life and can be classed as maintenance, the migrations that occur during regeneration and restitution following depletion after treatment with specific antisera, immuno-suppressive agents or irradiation which can be considered as regulation, and the migrations occurring after specific antigenic stimulation. In the first instance, the early embryogeny of the immune systems, cells from the yolk sac migrate to the embryonic liver or thymus to become the precursors of the B (bursal) type of lymphoid cell destined to form immunoglobulins, or the T (thymic) type of lymphoid cell slated to become a collaborator or "helper" in immunological reactions involving immunoglobulin synthesis, or "effector" cells as well as "helper" cells in cellular immune systems such as the graft-versus-host reaction [cf. 21, 22; cf. also 20]. To what extent specific antigenic properties of the cells are essential for the selective migration, or to what extent they result from it has not yet been determined, since marker studies with theta or TL antigens have not yet been applied to the earliest phases of lymphoid development [cf. 23].

A second set of cell migrations occurs after the differentiation of lymphoid cells has progressed in the "primary" lymphoid organs such as the thymus, liver or bone marrow. In this second set, cells already differentiated to T or B type, as judged by surface antigen markers [24, 25], leave these organs, and through the lymphatics, make their way to the "secondary" organs such as lymph node and spleen. This migration is also specific, as judged from restitution experiments involving infusion of labelled cells [26, 27], as well as by the fact that the rate and percentage distribution of cells is typical for the tissue of origin.

The migration pattern of cells involved in the regulation of lymphoid systems following lymphoid cell depletion has not been well mapped out, but presumably follows rules similar to those involved in the normal maintenance of lymphoid cell levels in the body. Aside from the intrinsic migration properties of cells, both in maintenance and restitution one needs to take into account the role played by such factors as lymphopoietin or thymic humoral components in release, or in the seeding of cells in the secondary lymphoid organs. To date the role of these humoral factors has not been sufficiently well defined to permit analysis.

Migration of cells following antigenic stimulation presents a final and perhaps most interesting phenomenon: upon antigenic stimulation there appears to be release of cells from bone marrow and thymus [28, 29], followed by selective sequestration of these cells in the spleen and other lymphoid effector organs. Why such a release of cells occurs is not known, and it is not clear whether the source of stimulation to release is antigen itself, antibody, or some specific antigen-antibody complex. At any rate, the selective nature of this release and cell migration process suggests that at least the activation of migration involves surface materials related to the immunoglobulin-antigen recognition system characteristic of immune systems. How this system is translated into a stimulus for release on the one hand and sequestration on the other is a problem whose solution seems imperative to an understanding of immunological function.

That *inductive tissue interactions* play a role in the ontogeny and function of immune systems has already been implied, and has been discussed extensively elsewhere [6, 15]. Inductive tissue interactions presumably are the basis for the elaboration of the initial lymphoid population in the thymus [30], and it is presumably at this time that the specific pathway of T versus B cells first becomes clearly established as divergent. In thymic inductive interaction, epithelial and mesenchymal components are both involved, which together influence the proliferation and differentiation of the lymphoid precursor stem cell, which itself is now believed to be of yolk sac origin [22]. The differentiation of lymphoid-like cells does not normally occur in the liver, but can be elicited *in vitro* [31] presumably again dependent on interaction between the precursor cells of yolk sac origin and the liver milieu of endodermal epithelium and mesodermal mesenchymal cells. It is of some interest here that the inductive systems seen to operate here are essentially analogous to the three-cell systems described for the development of the lens [cf. 32] and may presage the interaction of the adult functional immune system.

That such an adult functional system exists is suggested both by the three-cell interactive systems believed to operate in the eliciting of immune reactions *in vitro* [14], and by the regenerative systems demonstrated to operate following irradiation or urethan treatment [33]. In the former, interaction between macrophage-like adherent (mesenchymal) types of cells and non-adherent (lymphoid)

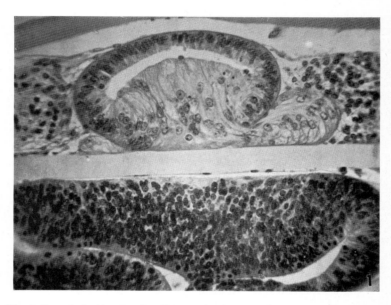

Fig. 1. Lens induction *in vitro*. Interaction between lens ectoderm (forming lens), mesenchyme (adjacent) and embryonic retina on opposite side of millipore filter (from MUTHUKKARUPPAN [32])

cell populations of T and B type has been suggested as essential to immunological function. In the latter instance, lymphoid depletion following either irradiation or urethan treatment of thymus can be overcome by inductive interaction between stem cells furnished by thymus or bone marrow and the appropriate inductively active tissues.

The uniquely sensitive selective interactive systems involved in this latter instance should be emphasized. For example, after urethan treatment, a thymic explant retains stem cells and connective tissue, but no lymphoid cell differentiation can take place. Bone marrow provides the inductive stimulus necessary for redifferentiation, and this influence follows the operational description of inductive tissue interaction, in that it is specific, can be mediated

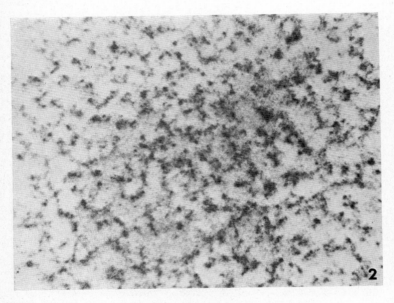

Fig. 2. Spleen suspension culture, 4 hours after explantation. Note early aggregates

across an intervening millipore filter barrier, and involves both differentiation and induction of cell division. Similarly, in immunological regeneration following irradiation, spleen, thymus and marrow must all interact, and here it has been shown that the stem cells derived from the bone marrow require both the splenic milieu and an inductive factor obtained from the thymus in order to redifferentiate to immunocompetence *in vitro* [34].

To what extent *aggregation of cells* plays a role distinct from that of cell migration as discussed above is not clear.

In *in vitro* systems involving explantation of spleen cell suspensions, immunological function can be demonstrated only when cell aggregation does, in fact, occur. Thus cell suspension systems must include cell types that can coaggregate and reconstruct follicles typical of intact spleens [15], and in the absence of cell aggregation

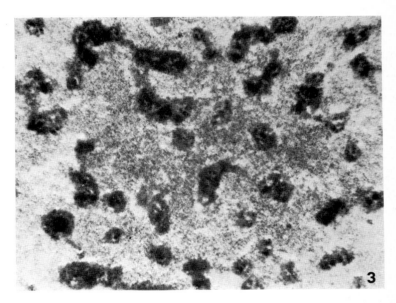

Fig. 3. Same culture, 24 hours after explantation. Follicles have formed. Loose cells are red blood cells added as antigen. Black and white print of color diapositive; original aggregates whitish against background of red erythrocytes

no suspensions have yet been shown to be capable of *in vitro* response to a new antigen.

That this *in vitro* aggregation system is not an artifact of the experiment but represents a real process involved in *in vivo* response as well is suggested by the selective aggregation pattern seen in irradiated animals reconstituted with thymic or lymph node cells. Aside from the characteristic migration patterns already mentioned, lymphocytes of diverse origin actually settle selectively in appropriate regions of the irradiated spleen [35], and immunological reconstitution requires at least the participation, and presumably the necessary selective segregation or aggregation, of diverse cell types including B and T cells. It is not demonstrated but appears likely then that the selective aggregation within the splenic mass

Table 1. *Cooperation between thoracic duct cells and teratoma cells in response to sheep red blood cells*

Exp.	TDL	Tumor	Both	Index
1	0	80	1380	17.3
2	25	26	80	1.6
3	20	27	192	4.1
4	0	7	80	11.4
5	130	10	760	5.4
6	10	55	530	8.2
7	867	10	1895	2.2
8	970	0	880	0.9
9	240	0	3447	14.4
n = 30		mean index = 5.78		

In each experiment, irradiated host animals were injected with 10^8 sheep red blood cells, and either thoracic duct lymphocytes (10^6), tumor cells (10^7) or both. Index of cooperation is calculated as plaque forming cells of doubly injected animals/plaque forming cells of the singly injected animals. In these experiments the cooperation index of bone marrow cells with thoracic duct lymphocytes averaged 2.0.

that is observed following lymphoid restitution represents a necessary component of that restitution.

In recent experiments in my laboratory we have been trying to put together the concepts of cellular interactions and communications as described in this paper, by attempting to control the entire differentiative sequence from initial embryonic stem cell to functional immunocompetence. To do this we have chosen embryonic tumor cells, or teratomas, using as source teratoma cells of line 129 (402C and 6050, Dr. LEROY STEVENS). Arguing that such teratoma cells, competent to differentiate into many different definitive cell types [36, 37], could become specifically precursors of lymphoid cells, hence presumptive T or B cells, we have injected tumor cells into lethally irradiated animals together with thoracic duct lymphocytes (known as "T" active cells) and antigen (sheep erythrocytes). Under these conditions we felt that tumor cells would selectively settle in the spleen, aggregate there in typespecific manner, interact inductively with the cells of the spleen and the T cells introduced simultaneously, and thereby differentiate to become antibody-producing cells.

These experiments, still in their infancy [38], suggest that at least superficially the tumor cells can participate in immunological reactions when tested in this manner. As can be seen from Table 1, animals reconstituted with tumor cells and thoracic duct lymphocytes can respond to sheep erythrocytes in a manner that animals restored with either TDLs or tumor cells alone cannot. Moreover, the cell collaboration between tumor and thoracic duct cells exceeds that seen in the now classical bone marrow and thoracic duct cells reconstitution system. The meaning of these experiments is not yet clear, for in the absence of surface markers or immunoglobulin analysis it is not yet known whether the tumor cells act specifically, and whether it is they or the thoracic duct cells which finally synthesize the observed antibody.

In conclusion, it is apparent that cell contacts and communications play an integral role in the total of differentiative processes associated with embryogeny and operative throughout adult life. The nature of these contacts and communications is doubtless varied, and the precise mechanisms whereby they affect differentiation is not yet understood. There is enough similarity between the various systems under study however to suggest that ultimately an understanding of these processes will help clarify not only the mechanism of differentiation in general, but specifically the control of events as seen in any individual differentiative system such as that of immunity.

References

1. BONNER, J. T.: The cellular slime molds, 2nd Ed. Princeton, N. J.: Princeton Univ. Press 1967.
2. AUERBACH, R.: The organization and reorganization of embryonic cells. In: Self-organizing systems. Oxford: Pergamon Press 1960, p. 101.
3. GROBSTEIN, C.: Differentiation of vertebrate cells. In: The Cell 1, 437 (1959) (BRACHET, J., MIRSKY, A. E., Eds.).

4. DAVIDSON, E. H.: Gene activity in early development. New York: Academic Press 1968.
5. AUSTIN, C. R.: Ultrastructure of fertilization. New York: Holt, Rinehart, Winston 1968
6. AUERBACH, R.: Toward a development theory of antibody formation: The germinal theory of immunity. In: Developmental aspects of antibody formation and structure (STERZL, J., RIHA, I., Eds.). New York: Academic Press 1971.
7. HOLTFRETER, J.: Arch. exp. Zellforsch. **23**, 169 (1939).
8. LILIEN, J. E.: Toward a molecular explanation for specific cell adhesion. Curr. Top. Develop. Biol. **4**, 169 (1969) (MONROY, A., MOSCONA, A. A., Eds.).
9. LILIEN, J.: Develop. Biol. **17**, 657 (1968).
10. MCQUIDDY, P., LILIEN, J. Cell. (1971) (in press).
11. GLOBERSON, A., AUERBACH, R.: Science **149**, 991 (1965).
12. — — J. exp. Med. **124**, 1001 (1966).
13. MISHELL, R. I., DUTTON, R. W.: Science **153**, 1005 (1966).
14. MOSIER, D. E.: J. exp. Med. **129**, 151 (1969).
15. AUERBACH, R.: Toward a developmental theory of immunity. In: Cell interactions and receptor antibodies in immune responses (MAKELA, O., CROSS, A., Eds.), p. 393. Academic Press (1971).
16. — Tissue interactions in vitro. In: Epithelial-mesenchymal interactions, p. 200 (FLEISCHMAJER, R., Ed.). Baltimore: Williams & Wilkins 1968.
17. — Continued inductive tissue interaction during differentiation of mouse embryonic rudiments in vitro. In: Retention of functional differentiation in cultured cells. Wistar Monogr. **1**, 3 (1964).
18. GROBSTEIN, C.: Science **143**, 643 (1964).
19. FLEISCHMAJER, R. (Ed.): Epithelial-mesenchymal interactions. Baltimore: Williams & Wilkins.
20. MAKELA, O., CROSS, A., KOSUNEN, T. U. (Eds.): Cell interactions and receptor antibodies in immune responses. New York: Academic Press 1971.
21. AUERBACH, R.: Develop. Biol. Suppl. **1**, 254 (1967).
22. MOORE, M. A. S., OWEN, J. J. T.: J. exp. Med. **126**, 715 (1967).
23. HAMMERLING, U.: This symposium.
24. SCHLESINGER, M.: Progr. exp. Tumor Res. (Basel) **13**, 229 (1970).
25. RAFF, M.: The use of surface antigenic markers to define different populations of lymphocytes in the mouse. In: Cell interactions and receptor antibodies in immune responses (MAKELA, O., CROSS, A., KOSUNEN, T. U. Eds.), p. 83. New York: Academic Press (1971).
26. GOWANS, J. L., KNIGHT, E. J.: Proc. roy. Soc. B **159**, 257 (1964).
27. FORD, C. E.: Traffic of lymphoid cells in the body. In: CIBA Foundation Symposium: The Thymus,p. 131. (WOLSTENHOLME, G. E. W., PORTER, R., Eds.). London: Churchill 1966.
28. ABDOU, N. I., RICHTER, M.: Int. Arch. Allergy **35**, 330 (1969).
29. SPRENT, J., MILLER, J. F. A. P., MITCHELL, G. F.: Cell Imm. **2**, 171 (1971).
30. AUERBACH, R.: Develop. Biol. **3**, 336 (1961).

31. UMIEL, T., GLOBERSON, A., AUERBACH, R.: Proc. Soc. exp. Biol. (N. Y.) **129**, 598 (1968).
32. MUTHUKKARUPPAN, V.: J. exp. Zool. **159**, 269 (1965).
33. GLOBERSON, A., AUERBACH, R.: In vitro studies on thymus and lung differentiation following urethan treatment. In: Methodological approaches to the study of leukemias (DEFENDI, V., Ed.). Wist. Inst. Press 1965.
34. — — J. exp. Med. **126**, 223 (1967).
35. CONGDON, C. C., MAKINODAN, T., GENGOZIAN, N., SHEKARCHI, I. C., URSO, I. S.: J. nat. Cancer Inst. **21**, 193 (1958).
36. PIERCE, G. B., DIXON, F. J., VERNEY, E. L.: Lab. Invest. **9**, 583 (1960).
37. STEVENS, L. C.: Develop. Biol. **2**, 285 (1960).
38. AUERBACH, R.: Proc. 1st Int. Congr. Cell Diff., Abstract, 1971 (in press).
39. — Proc. Vth Int. Symp. Comp. Leukemia Research (CHIECO-BIANCHI, DUTCHER, R. M., Eds.). Basel: Karger 1971.

Some Aspects of the Dynamics of Cell Surface Antigens

ULRICH HÄMMERLING

Institut für Virologie, Justus Liebig-Universität, 63 Giessen, West Germany

With 1 Figure

The main topic of the 22nd Colloquium of the "Gesellschaft für Biologische Chemie" being the dynamics of cellular membranes, I should like to review some genetic aspects of surface antigens, with emphasis on those phenomena that may have some bearing on the dynamics of cell surfaces. When speaking of dynamics I am not referring to the biochemical turnover of surface components, but to the phenotypic changes in general that a somatic cell undergoes during its life span. It has been realized that surface antigens do not represent stable, unalterable phenotypic traits. In the following, I should like to discuss some factors responsible for such phenotypic changes.

The phenomena that we shall consider all refer to components of the cell that are identified by immunological methods and therefore are called antigens. This does not imply any other function. The physiological role of these components, although unknown, must in the majority of cases be other than immunological.

Perhaps we can take a systematic approach, and deal with the following phenomena:

1) Changes of cell surface composition by external factors, i.e. factors not directly controlled by the cell.

2) Acquisition and loss of genetic information leading to modifications in cell surface structure, and

3) Repression or derepression of cellular genes in differentiating cells causing phenotypic surface changes.

Re 1. External factors may cause the *acquisition of new antigens*, as well as the disappearance of cell-coded antigens. The most trivial example of acquisition of an antigen is passive adsorption.

Well-known examples in serology are the adsorption of bacterial lipopolysaccharides [1], or cardiolipin on red blood cells, which form the basis of a variety of diagnostic tests with a wide range of application. Passive adsorption is not an altogether unphysiological process; it is also observed *in vivo*. For example, the J blood group substance in cattle is a glycolipid occurring in free form in plasma, and by adsorbing to red cells it adds a new antigenic specificity to the existing ones [2]. Other examples illustrating that passive adsorption is probably a general event *in vivo* are Gross soluble antigen [3] that is taken up by normal lymphocytes, and cytophilic antibodies, that modify monocytes and macrophages [4].

The reverse of antigen acquisition, namely the *phenotypic loss of cellular antigens*, is also known. This phenomenon is exemplified by antigenic modulation [5], the prototype of which concerns the TL alloantigen of mouse thymocytes. The decisive factor here is the specific antibody against TL. When thymus cells are exposed to anti-TL antibody *in vitro* or *in vivo* (i.e. by passive immunization of the animal), within a period of time the antigen is lost from the surface. TL is resynthesized when the cell is returned to its normal, antibody-free milieu.

It has been demonstrated that TL modulation does not involve a simple steric coverage of the antigenic sites by antibody, but that TL is in fact lost from the surface by an as yet unknown mechanism. In modulated cells it is neither possible to produce lysis by addition of complement nor can the originally present TL antibody be demonstrated by fluorescent anti-mouse gamma globulin. In addition, antisera against subspecificities such as TL-1 induce modulation of the entire TL complex, including the TL specificities 2 and 3 situated on the same molecule. The fate of TL, whether it is ingested by the cell or whether it is exfoliated into the milieu, is not known.

TL modulation has an important biological consequence: TL is also found on leukemias of mice, including those strains that normally do not express TL on their thymocytes. These TL-negative strains are capable of producing antibody against TL. However, a TL-positive tumor will grow in syngeneic TL-negative strains, as by modulation the tumor cells are capable of escaping the immunological attack the host is mounting. Therefore, TL is not a transplantation antigen.

Modulation, i.e. loss of surface constituents from the cell through the action of antibody, is also observed in other antigenic systems. A recent example is the surface-associated immunoglobulin on mouse lymphocytes. TAKAHASHI [6] found that rabbit antibody to kappa chains in the presence of complement is cytotoxic for a proportion of mouse lymphocytes. But when the cells are incubated with anti-kappa alone and C^1 is added later, no lysis occurs. Thus, there is a striking similarity with TL modulation. TAKAHASHI has also demonstrated that H-2 antigen can be removed to some extent from the surface in a process resembling modulation [6]. In this case, however, an antibody to mouse globulin is needed in addition to H-2 antibody, in a type of piggy-back experiment leading to about 30% reduction of H-2. Other antigens can be expected to modulate, too, and it is possible that cells use modulation as a repair mechanism to replace blocked-up parts of their surface.

Re 2. *Genotypic acquisition of surface antigens.* This includes phenomena that depend on the transfer of nucleic acids, either RNA or DNA. The process has also been termed antigenic conversion — which is a rather unfortunate term because conversion implies the chemical modification of cellular antigens (for ref. see the conversion of bacterial polysaccharides by lysogenic phages [7]), in contrast to the expression of new components coded for by extraneously acquired genetic information. The examples that we will consider include the infection of somatic cells with RNA or DNA viruses.

New surface antigens occur as a consequence of virus replication of RNA as well as DNA viruses. It is often difficult to decide whether these components are virus- or cell-coded, so that two mechanisms can be held responsible: a) incorporation into the plasma membrane of virus-coded products that are not necessarily components of the virion and b) control by virus of the disposition of cellular antigens. According to ROIZMAN [8], herpes virus codes for the peptide chain of glycoproteins which are associated with the surface as well as internal membranes of the host cell. RNA viruses, too, code for components that are incorporated into the surface and are recognized as antigens [9]. To take an example studied by AOKI [10]: infection with murine leukemia virus (GROSS) leads to the appearance of a Gross-specific antigen, the G-antigen, occurring on the cell membrane only, and not on the budding virus particle. Examples of surface components that are dependent on the synthesis of virus itself are numerous. Dr. KLENK in the next presentation will discuss these problems in more detail [11].

We shall now consider the alternative, that virus controls the disposition of cellular antigens. Fox et al. [12, 13] recently demonstrated that 3T3 cells transformed by polyoma virus exhibit a surface component which serves as a receptor for wheat germ agglutinin [14—16], and which normally is absent from the cells except for cells in mitotic division. It appears that the virus genome influences the membrane structure and controls the expression or the accessibility of the receptors without coding for them. The components are synthesized in normal cells because mild treatment with trypsin uncovers the wheat germ receptors. In transformed cells, however, the receptors are accessible in all phases of the cell cycle. Another interpretation recently put forward by NICOLSON [17] is that in normal cells the regional concentration of receptors is too low to allow agglutination; in contrast the surface organization is altered in transformed or trypsin-treated cells in such a way that receptors are more concentrated in certain areas, facilitating agglutination. The number of receptor sites may be less important than the local concentration in parts of the cell surface.

Although there is at present no clear-cut example of an antigen or other component as a direct product of an integrated viral gene, this possibility is most interesting. Covalent linkage to chromosomes has indeed been demonstrated for DNA viruses. DOERFLER [18] by sophisticated hybridization techniques showed the integration of at least part of the adeno 12 genome, and according to ZUR HAUSEN [19] there is evidence for a close chromosomal association of the EBV genome in certain BURKITT cell lines. In the case of SV40, a product of the integrated gene seems to be demonstrable as a transplantation antigen [20]. The final proof that this is indeed a virus-coded component and not a product of a derepressed cellular gene remains to be sought. The rigorous tests of formal genetics, segregation etc., required to prove linkage are at present not feasible with somatic cell lines. But it would be difficult to explain why transformed cell lines from different species should form antigens of identical specificity.

Whereas there is a rationale for the integration of DNA viruses, the mechanism of RNA virus integration is less readily comprehended. The reverse transcriptase of TEMIN [21] and BALTIMORE

[22], originally found in Rous sarcoma virus but subsequently also found in other oncogenic RNA viruses, may offer a basis for integration. The reverse transcriptase may give a clue as to why oncogenic viruses, such as MuLV (murine leukemia virus) in high-incidence strains of mice, or mammary tumor virus (MTV) in the GR strain (see BENTVELZEN [23]), are transmitted vertically through the gametes, and why certain products of oncornaviruses such as the GS antigen of ALV (avian leucosis virus), behave like dominant Mendelian traits. The most likely form of such a close association with the gametes would be chromosomal integration similar to lysogeny in bacteria. BOYSE, OLD and STOCKERT [24, 25] have suggestive evidence that at least part of the MuLV genome may be integrated into the genome of some strains of mice. The hypothesis is based on the finding that the G_{IX} antigen, a surface antigen of thymocytes of normal mice that is possibly coded by Gross virus, is genetically linked with H-2 in the 9th linkage group.

The examples we have just discussed were meant to illustrate that the phenotype of somatic cells can be directly influenced by the translation of viral information, or indirectly by repression of cellular genes, or by controlling the disposition of membrane components. Finally, I would like to mention, that changes in phenotype may also be induced by DNA extracts. In BORENFREUND and BENDICH's experiment [26], which is a repetition of the classical AVERY experiment [27], a DNA extract of Ehrlich ascites cells induced surface antigens in Chinese hamster cells specific for the DNA donor (EHRLICH).

Loss of a surface antigen in connection with deletion of the corresponding gene — the contrary of what we have just discussed — is also observed. This phenomenon differs from modulation where only the phenotype is affected and the structural gene is preserved.

BJARING et al. [28] review the production of what are called "parental variants" of transplanted tumors: tumors, originating in F_1 hybrid (H-2 heterozygous) mice and therefore possessing both sets of parental antigens, are in general not transplantable in the parental strains. On repeated passage, however, in congenic strains histoincompatible in H-2 antigen only, it is possible to isolate cell lines that have irreversibly lost the histoincompatible antigen, as a result of mutation. It was demonstrated that only one allele is deleted and that the deletion affects the neighboring loci as well. This experimental system deserves attention because it may open a way of studying genetic linkage in somatic cell lines.

3) Phenotypic changes occurring as a consequence of *repression or derepression of cellular genes.*

In the following I should like to compare the antigenic structure of mouse lymphoid cells and attempt to draw conclusions as to the genetic events occurring in differentiating cells. This approach is justified, as on mouse lymphocytes there are a number of surface antigens known which represent most probably primary gene products. There is suggestive evidence [29] that in H-2 the antigenic determinants form part of the polypeptide chains, and the surface-associated immunoglobulin is another example of a primary gene product.

In other antigenic systems, e.g. the TL alloantigens, the Ly group of antigens, and Θ, the relationship to the structural genes, whether these antigens represent primary or secondary gene products, is less well resolved, but in so far as they are macromolecules they may nevertheless be used as markers for the cell surface. With the exception of H-2 and Θ, the above antigens are restricted to lymphoid cells, but it is apparent that, even within the family of lymphocytes, there is widespread antigenic variation, both with regard to quantity and to quality of antigen expression. We are coming to recognize certain trends in lymphoid cells suggesting that expression of many antigens is dependent on selective gene action. Thus the term differentiation antigen seems appropriate [30].

The surface antigens exhibit vast polymorphism within the species, and it is indeed this polymorphism that enables the investigator to analyse them by immunological methods and to perform the kinds of serological and genetical tests that have rendered immunogenetics a powerful approach to study differentiation. Because of the availability of these immunogenetic techniques, which will be discussed below, the following questions can be asked realistically:

1) Which are the cells that carry certain antigens, and where is the exact location on the surface?

2) What are the quantitative relationships?

3) How do the antigen patterns change with progressing cell maturation?

An important technical advance in serology has been the introduction of ferritin-labeling, allowing us to localize antigens on individual cells [31]. This extended serological approach has become feasible since the introduction of hybrid antibodies for visual labeling [32, 33]. In principle, labeling by hybrid antibodies is a complex but highly specific way of attaching ferritin or another marker to antibody and producing sufficient electron density to visualize the position of antibody molecules in the electron microscope. By applying this technique to murine cells, new information concerning the distribution of alloantigens on the surface has been obtained. The general observation was that alloantigens are not uniformly distributed but are restricted to certain areas and are lacking in other parts of the cell surface [31]. This was observed not only for H-2 on lymphocytes but also for other alloantigens, and for other mouse cells [31]. Human cells were also found to exhibit a discontinuous distribution of HL-A [34]. This in fact is what was suggested by earlier evidence with the immunofluorescence method [35].

Different alloantigenic systems exhibit contrasting regional distribution on different cells. For instance, on thymocytes, H-2 occupies only small areas, amounting to about 5% of the surface; on the same cells, however, Θ covers more than 50%, and Ly-A, too, is represented on a substantial portion of the surface. A peripheral lymphocyte shows an almost reciprocal distribution: H-2 is found to cover about 50% of the cell perimeter, whereas Θ and Ly-A are reduced to about 5 to 10% of the surface.

By employing tritium-labeled antibodies to alloantigens we have estimated the number of available antigenic sites per cell [36]. Although the number of alloantibody molecules bound per cell that is measured in this test does not equal the number of antigen molecules, it represents a close approximation. There is an exact parallel between the proportion of surface area covered and the quantity of alloantigen per cell. For instance, in the electron microscope H-2 was found to reside in densely packed sectors covering about half the perimeter of lymph node lymphocytes, in contrast to about 5 to 10% of the thymocyte perimeter. This corresponds well with the quantity of 600,000 anti-H-2 molecules bound per lymphocyte versus 100,000 anti-H-2 molecules on thymocytes. Conversely, Θ is seen to occupy the majority of the thymocate surface whereas only small sectors are found on lymph node lymphocytes. This again is supported by quantitative measurements: a thymocyte binds about 400,000 anti-Θ molecules whereas lymph node lymphocytes on the average bind only 70,000 anti-Θ molecules. The latter figure cannot be readily interpreted because lymph node populations contain a proportion of B-lymphocytes that do not express Θ. However, the difference is too large to be explained by dilution of T-lymphocytes with B-lymphocytes (30 to 40%) and more likely reflects a true quantitative difference in antigen expression of T-lymphocytes as compared to thymocytes. Thus, as a general rule, the quantity of antigen available for absorption of antibody may be determined by the relative size of the sectors where the antigen is represented, and not by the density of antigen. Within the areas of antigen representation the density may well be constant.

The blocking test devised by BOYSE et al. [37] suggests that the arrangement of surface antigens is not random. This test takes advantage of the fact that an antibody molecule requires a space larger than the determinant group against which it is directed, and therefore may sterically cover another site close by. Because antigens are fixed in the two-dimensional lattice of the surface membrane, the neighboring sites may even form parts of different molecules.

In essence the information obtained by the blocking test shows that some pairs of alloantigens reside in close proximity because there is sterical interference in the attachment of the corresponding antibodies. Other pairs must reside in discrete areas as there is no competition for antibody absorption. It was concluded that antigens may be allocated in relation to each other according to a prescribed pattern.

The concept emerging from immunoferritin studies can be appropriately described as regional differentiation of the cell surface. At present, this is a purely morphological description, neither its functional significance nor its origin being known. It is possible that the regional amassing of antigens, whatever their physiological role, imparts a synergistic advantage to the cell. As to the origin of regional differentiation, we surmise that it reflects in some way a characteristic mode of membrane assembly. Antigenic components may be exchanged as individual units in the course of a general biochemical turnover, but as a more attractive alternative, that would obviate the necessity of a rigid lattice structure, antigenic components may be incorporated into newly assembled membrane in a mechanism resembling crystallization. Such a model of membrane assembly would easily explain the allocation in the areas of predominant antigen representation that we observe by immune electron microscopy. Furthermore, during morphogenesis varying sets of genes become derepressed, and the products of such genes include surface antigens ("differentiation antigens"). Under the assumption that there is a corresponding mechanism of gene regulation, the regional differentiation observed can be understood in a dynamic process, as a constant flow of membrane components over the surface from one or more centers of growth, adapting the cell to its varying physiological needs.

There is initial evidence to substantiate the hypothesis that differentiation of lymphocytes is accompanied by extensive changes in antigenic surface structure. With the reservation that the derivation of lymphocytes is still controversial, which is an inevitable complication when working with cell populations, a comparison of the phenotypes of thymocytes and peripheral lymphocytes may furnish some clues as to the pathway of differentiation. In order to assess the major changes in surface structure, it may be necessary to consider qualitative as well as quantitative differences in antigen profiles. During maturation of thymocytes into lymphocytes the following events can be assumed to take place in what I believe is a dynamic process (Fig. 1): while the thymus cells reside in the thymus they acquire the TL antigen which is most probably not represented on the bone marrow stem cell [38]. TL expression may be transient and cease while the cells are still within the thymus, because a subpopulation of thymocytes can be distinguished that are TL-negative but have already acquired graft-versus-host activity [39]. In contrast, the alloantigen Θ and the Ly group continue to be expressed after the cells have emerged from the thymus, although the synthesis is greatly diminished (for Θ by a factor of about five).

The opposite trend of an increase in antigen representation with progressing maturation is exemplified by the alloantigen H-2 that occurs in a 6-fold higher concentration on lymphocytes as compared to thymocytes, and by the surface-associated immunoglobulin (Ig)

[40]. The latter according to electron microscopic data appears to be represented on 90% of mature lymphocytes from lymph nodes and therefore is a component of T- as well as B-lymphocytes [40]. It is most intriguing that Ig is virtually absent from thymocytes, and may not be acquired before the cell is sufficiently mature to leave the thymus or when it has arrived in the periphery.

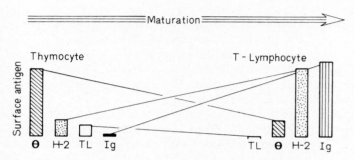

Fig. 1. The maturation of T-lymphocytes is characterized by changing profiles of surface antigens involving quantitative as well as qualitative alterations. The comparison of the antigen profiles, based on studies with ferritin- and tritium-labelled antibodies, of thymus and lymph node cell populations reveals considerable diversity. The antigen profiles in the diagram are expressed as the approximate numbers of antigen molecules per cell and correspond to the surface area covered. The differences cannot be explained by the presence in lymph node populations of B-lymphocytes but more likely reflect the dynamic changes in surface composition occurring during normal differentiation from thymocytes into lymphocytes

The above scheme of T-lymphocyte maturation is still hypothetical, but it may form a basis to distinguish morphologically between the two major subdivisions of lymphocytes [41], namely the B-lymphocytes, which are responsible for the production of humoral antibody, and the T-lymphocytes, which mediate the cell-associated immunity, and even between subpopulations within T-lymphocytes. A corresponding scheme of differentiation antigens for B-lymphocytes has as yet to be devised. The reason here is that in B-lymphocytes we are lacking a homogeneous population of precursor cells such as the thymocytes provide for T-lymphocytes. With the discovery of MBLA by RAFF et al. [42] and the PC-1 alloantigen by TAKAHASHI [43] occurring on normal and malignant plasma cells and possibly representing a "late" B-cell antigen, this gap begins to fill. Our incomplete inventory of the antigens of lymphoid cells can be summarized in the following table. As a result of the analysis of antigens of lymphoid cells we begin to realize that the surface phenotypes of cells are changing in a systematic fashion during differentiation. Aside from the practical value of using antigen profiles for delineating lymphocyte differen-

Table 1. *Surface antigens of immunocytes*

	Ig[a]	H-2	TL	Θ	Ly	MSLA[b]	PC-1	MBLA
thymocytes	—	+	+ +	+ +	+ +	+ +	—	—
T-lymphocytes	+ +	+ +	—	+	+	+	—	—
B-lymphocytes	+ +	+ +	—	—	—	?	+	+
plasma cells	—	+ +	—	—	—	?	+ +	+

[a] surface-associated immunoglobulin [40].
[b] MSLA: mouse specific lymphocyte antigen [44].

tiation, this may be regarded as a special instance of the more general rule that phenotypic differences among somatic cells are the reflections of selective gene action.

Acknowledgement

I am indebted to Professor H. J. EGGERS, Giessen, for his continuous interest in the work, and to Dr. E. A. BOYSE, New York, for many helpful discussions.

References

1. NETER, E., WESTPHAL, O., LÜDERITZ, O., GORZINSKY, E. A., EICHENBERGER, E.: Studies of enterobacterial lipopolysaccharides. Effects of heat and chemicals on erythrocyte modifying antigenic, toxic and pyrogenic properties. J. Immunol. **76**, 377 (1956).
2. STONE, W. H., IRWIN, M. R.: Blood groups in animals other than man. Advanc. Immunol. **3**, 315 (1963).
3. AOKI, T., BOYSE, E. A., OLD, L. J.: Wild-type Gross leukemia virus. I. Soluble antigen in the plasma and tissues of infected mice. J. nat. Cancer Inst. **41**, 93 (1968).

4. HUBER, H., POLLEY, M. J., LINSCOTT, W. D., FUDENBERG, H. H., MÜLLER-EBERHARD, H. J.: Human monocytes: Distinct receptor sites for the third component of complement and for immunoglobulin G. Science 162, 1281 (1968).
5. OLD, L. J., STOCKERT, E., BOYSE, E. A., KIM, J. H.: Antigenic modulation: Loss of TL antigen from cells exposed to TL antibody. Study of the phenomenon in vitro. J. exp. Med. 127, 525 (1968).
6. TAKAHASHI, T.: Possible examples of antigenic modulation affecting H-2 antigens and surface immunoglobulins. Transpl. Proc., (1971) (in press).
7. LURIA, S. E., DARNELL, J. E.: In: General Virology, 2nd edition, p. 278. New York: J. Wiley and Sons, Inc. 1967.
8. ROIZMAN, B., SPRING, S. B.: Alteration in immunologic specificity of cells infected with cytolytic viruses. In: Crossreacting antigens and neoantigens (TRENTIN, J. F., Ed.). Baltimore: Williams and Wilkins Co. 1967.
9. SCHOLTISSEK, C., DRZENIEK, R., ROTT, R.: Myxoviruses. In: The biochemistry of viruses (LEVY, H. B., Ed.). New York, London: M. Dekker 1969.
10. AOKI, T., BOYSE, E. A., OLD, L. J., DE HARVEN, E., HÄMMERLING, U., WOOD, H. A.: G (Gross) and H-2 cell surface antigens: Location on Gross leukemia cells by electron microscopy with visually labelled antibody. Proc. nat. Acad. Sci. (Wash.) 65, 569 (1970).
11. KLENK, H. D.: Structure and biosynthesis of viral membranes. These proceedings, 1971.
12. BURGER, M. M.: A diffeience in the architecture of the surface membranes of normal and virally transformed cells. Proc. nat. Acad. Sci. (Wash.) 62, 994 (1969).
13. FOX, T., SHEPPARD, J. R., BURGER, M. M.: Cyclic membrane changes in animal cells: Transformed cells permanently display a surface architecture detected in normal cells only during mitosis. Proc. nat. Acad. Sci. (Wash.) 68, 244 (1971).
14. INBAR, M., SACHS, L.: Structural difference in sites on the surface membrane of normal and transformed cells. Nature (Lond.) 223, 710 (1969).
15. — — Structural difference in sites on the surface membrane of normal and transformed cells. Nature (Lond.) 223, 710 (1969).
16. SACHS, L., INBAR, M.: Int. J. Cancer 4, 690 (1969).
17. NICOLSON, G.: Pers. communication, 1971.
18. DOERFLER, W.: Integration of the deoxyribonucleic acid of adenovirus type 12 into the deoxyribonucleic acid of baby hamster kidney cells. J. Virol. 6, 652 (1970).
19. ZUR HAUSEN, H., SCHULTE-HOLTHAUSEN, H., KLEIN, G., HFNLE, W., HENLE, G., CLIFFORD, P., SANTESSON, L.: EBV DNA in biopsies of Burkitt tumours and anaplastic carcinomas of the nasopharynx. Nature (Lond.) 228, 1056 (1970).
20. DULBECCO, R.: Cell transformation by viruses. Science 166, 962 (1969).
21. TEMIN, H. T., MIZUTANI, S.: RNA-dependent DNA polymerase in virions of Rous sarcoma virus. Nature (Lond.) 226, 1211 (1970).

22. BALTIMORE, D.: RNA-dependent DNA-polymerase in virions of RNA tumour viruses. Nature (Lond.) **226**, 1209 (1970).
23. BENTVELZEN, P., DAAMS, J. H., HAGEMANN, P., CALAFAT, J.: Genetic transmission of viruses that incite mammary tumor in mice. Proc. nat. Acad. Sci. (Wash.) **67**, 377 (1970).
24. STOCKERT, E., BOYSE, E. A., OLD, L. J.: The G_{1x} System: A cell surface allo-antigen associated with murine leukemia virus; implications regarding chromosomal integration of the viral genome. J. exp. Med. **133** 1334 (1971).
25. BOYSE, E. A., OLD, L. J., STOCKERT, E.: The relation of linkage group IX to leukemogenesis in the mouse. In: Proceedings of the Conference on RNA. Viruses and Host Genome in Oncogenesis (EMMELOT, P., BENTVELZEN, P., Eds.). Amsterdam: North-Holland Publishing Co., 1971 (in press).
26. BORENFREUND, E., HONDA, Y., STEINGLASS, M., BENDICH, A.: Studies of DNA induced in heritable alteration of mammalian cells. J. exp. Med. **132**, 1071 (1970).
27. AVERY, O. T., MACLEOD, C. M., MCCARTY, M.: Studies of the chemical nature of the substance inducing transformation of pneumococcal types. J. exp. Med. **79**, 137 (1944).
28. BJARING, B., KLEIN, G.: Antigenic characterization of heterozygous mouse lymphomas after immunoselection *in vivo*. J. nat. Cancer Inst. **41**, 1411 (1968).
29. MURAMATSU, T., NATHENSON, S.: Studies on the carbohydrate portion of membrane-located mouse H-2 alloantigens. Biochemistry **9**, 4875 (1970).
30. BOYSE, E. A., OLD, L. J.: Some aspects of normal and abnormal cell surface genetics. Ann. Rev. Genet. **3**, 269 (1969).
31. AOKI, T., HÄMMERLING, U., DE HARVEN, E., BOYSE, E. A., OLD, L. J.: Antigenic structure of cell surfaces. An immunoferritin study of the occurrence and topography of H-2, Θ, and TL alloantigens on mouse cells. J. exp. Med. **130**, 979 (1969).
32. HÄMMERLING, U., AOKI, T., DEHARVEN, E., BOYSE, E. A., OLD, L. J.: Use of hybrid antibody with anti-γG and anti-ferritin specificities in locating cell surface antigens by electron microscopy, 1968.
33. — —, WOOD, H. A., OLD, L. J., BOYSE, E. A., DEHARVEN, E.: New visual markers of antibody for electron microscopy. Nature (Lond.) **223**, 1158 (1969).
34. SILVESTRE, D., KOURILSKY, F. M., NICCOLAI, M. G., LEVY, J. P.: Presence of HL-A antigens on human reticulocytes as demonstrated by electron microscopy. Nature (Lond.) **228**, 67 (1970).
35. CERROTINI, J. C., BRUNNER, K. T.: Localization of mouse isoantigens on the cell surface as revealed by immunofluorescence. Immunology **13**, 395 (1967).
36. HÄMMERLING, U., EGGERS, H. J.: Quantitative Measurement of uptake of antibody on mouse lymphocytes. Europ. J. Biochem. **17**, 95 (1970).

37. Boyse, E. A., Old, L. J., Stockert, E.: An approach to the mapping of antigens on the cell surface. Proc. nat. Acad. Sci. (Wash.) **60**, 886 (1968).
38. Schlesinger, M., Hurwitz, D.: Serological analysis of the thymus and spleen grafts. J. exp. Med. **127**, 1127 (1968).
39. Leckband, E., Boyse, E. A.: Immunocompetent cells among mouse lymphocytes: A minor subpopulation. Science **172**, 1258 (1971).
40. Hämmerling, U., Rajewski, K.: Europ. J. Immunol. (1971) (in preparation).
41. Roitt, T. M., Greaves, M. F., Torrigiani, G., Brostoff, J., Playfair, J. H. L.: The cellular basis of immunological responses. Lancet **1969 II**, 367.
42. Raff, M. C., Nase, S., Mitchison, N. A.: Mouse specific bone marrow-derived lymphocyte antigen as a marker for thymus-independent lymphocytes. Nature (Lond.) **230**, 50 (1971).
43. Takahashi, T., Old, L. J., Boyse, E. A.: Surface alloantigens of plasma cells. J. exp. Med. **131**, 1325 (1970).
44. Shigeno, N., Hämmerling, U., Arpels, C., Boyse, E. A., Old, L. J.: Preparation of lymphocyte-specific antibody from antilymphocyte serum. Lancet **1968, II**, 320.

Glycolipid Changes Associated with Malignant Transformation*

SEN-ITIROH HAKOMORI

Department of Pathobiology, University of Washington, Seattle, WA 98105 USA

With 8 Figures

Introduction: General Contrasting Properties of Normal and tumor Cell Surfaces and Glycolipid Profiles of Cell Membrane

A major change in the cellular regulatory mechanism that is produced by the transformation of normal cells by carcinogenic agents can be ascribed to a change in the cell surface membrane [see 1]. As is shown in Table 1, (item a), the change in surface function of tumor cells is characterized by the loss of several properties of normal cell surfaces, such as contact inhibition [2], specific cell adhesion [3] and intercellular contact communication [4].

A loss of differentiated antigen (e.g. bloodgroup A and B antigens) with a simultaneous increase of less differentiated antigens (e.g. bloodgroup H or Lewis antigens), as observed in human epithelial tumors [5—11], is also noticeable (see Table 1, item b). Similarly, the presence of "less differentiated" antigens in human epithelial tumors may be essentially related to the presence of the "carcinoembryonic antigen" of GOLD [12, 13] or of an "autoantigen" of BURTIN et al. [14, 15]. Both antigens are glycoproteins, and are found in human gastrointestinal epithelial tumors as well as in the corresponding fetal organs, nevertheless they are reported to be "autoantigenic"; a similar retrogenic antigen has also been found in experimental tumors [16]. The presence of incompatible bloodgroup antigens, such as "A-like" antigen in tumors of host "B", or host "0" [11, 17, 18], and Lewis b antigen in tumors of

---

* Supported by National Cancer Institute grant (CA 10909) of the U.S. Public Health Services, and by the American Cancer Society Research Grant (T-475).

host "Le^a" [19], has been reported. Although the presence of tumor-specific surface antigen is well established [20], its relation to the retrogenetic antigen or to the incompatible bloodgroup antigens is still unknown. Presence of heterophilic antigen (Forssman

Table 1. *Contrasting Properties of Normal and Tumor Cell Surfaces*

	Normal	Tumor	References
a) Biological			
Contact Inhibition	+	—	2
(of locomotion and replication)			
Specific Intercellular Adhesion	+	—	3
Contact Communication	+	—	4
b) Immunological			
Differentiated Antigen	+	—	5—8
Less-Differentiated Antigen	—	+	9—16
(Retrogenetic antigen)			
Incompatible bloodgroup antigen	—	+	11, 17—19
Heterophilic antigen changes	—	+	21—23
Tumor-specific transplantation antigen	—	+	20
c) Physical and Chemical			
Negative surface charge	Low	High	24—26
New membrane protein component	—	+	36
Membrane heteroglycan changes			
i) incomplete carbohydrate chain synthesis			
glycolipid	—	+	28—36
glycoprotein	—	+	37—38
ii) Accumulation of precursor carbohydrate	—	+	28, 33, 35
iii) New glycopeptide component	—	+	39—40
Translocation of Carbohydrate sites			
(organizational change)			
i) Translocation; outwards	—	+	29, 41—44
ii) Translocation; inwards	—	+	45

type) in tumor and absence of it in the progenitor cells has been known in some experimental tumors and in *in vitro* transformed cells [21, 22, 23].

An increase of negative surface charge of cells demonstrated by cell electrophoresis [24, 25] or by histochemical staining [26] has been described as a surface change of various malignant cells (see Table 1, item c). The increased negative surface charge was initially ascribed to an increase in sialic acid [24, 25]; however, the increased quantity of sialic acid has not been supported by other investigators [27]. In our investigation, the total lipid-bound sialic acid of polyoma transformed "BHK" cells was very low as compared to normal "BHK" cells [28], whereas the total lipid-bound sialic acid of minimal-deviation Morris hepatoma was very high as compared to normal liver tissue due to an enormous accumulation of disialoganglioside [33, 34] which probably occurred as the result of blockage of the synthesis of trisialoganglioside (see section 5). The change in the total sialic acid content therefore seems to be closely dependent upon the blockage site in the synthetic route of carbohydrate chains containing sialic acid, so that an enormous variation of total sialic acid content could result.

The last two items in Table 1 concerning alteration of membrane heteroglycans and translocation of membrane carbohydrate sites could be the most general phenomena of malignancy, and I would like to discuss them extensively in this article. The phenomena listed in Table 1 may be interrelated and it is probable that some of these are different manifestations of the same process.

Before getting into a discussion on glycolipid changes on malignant transformation, it might be helpful to review the structure and organizational arrangement of glycolipid molecules in membranes. During the last decade, the chemistry of glycosphingolipids has been greatly advanced and the structure of major components has been disclosed after the pioneer work of KLENK, KUHN, SVENNERHOLM, WIEGANDT, and YAMAKAWA. Structurally related glycolipids can be classified in to five series of compounds: the analogues of ganglioside, globoside, hematoside and bloodgroup glycolipid, in addition to a basic structural unit of glycolipid (see Table 2).

Recently, the anomeric structure of ceramide trihexoside and globoside has been determined [49] and the structure of FORSSMAN hapten has been revised [50] as shown in Table 2. The distribution of these glycolipids in various parenchymatous, epithelial, and mesenchymatous cells varies markedly. It has been generally believed that the quantity and the quality of these glycolipids are genetically determined, and this belief is consistent with the analytical results of glycolipid composition of erythrocytes [48]. This

Table 2. *Glycolipids of animal cells*

1. Simple basic structure	Gal $\overset{\beta}{\rightarrow}$ Cer	Cerebroside
	Gal → Gal → Cer	CMH
	Glu → Cer	CMH
	Gal 1 → 4 Glu → Cer	CDH
2. Sulfatides	HSO_3 → 3 Gal → Cer	Sulfatide
	HSO_3 → 3 Gal 1 → 4 Glu → Cer	
3. Globoside series	Gal 1 $\overset{\alpha}{\rightarrow}$ 4 Gal 1 $\overset{\beta}{\rightarrow}$ 4 Glu → Cer	CTH
	GalNAc 1 $\overset{\beta}{\rightarrow}$ 3 Gal 1 $\overset{\alpha}{\rightarrow}$ 4 Gal 1 $\overset{\beta}{\rightarrow}$ 4 Glu → Cer	Globoside
	GalNAc 1 $\overset{\alpha}{\rightarrow}$ 3 GalNAc 1 $\overset{\beta}{\rightarrow}$ 3 Gal 1 $\overset{\alpha}{\rightarrow}$ 4 Gal 1 $\overset{\beta}{\rightarrow}$ 4 Glu-Cer	FORSSMAN
4. Hematoside series	NANA 2 $\overset{\alpha}{\rightarrow}$ 3 Gal 1 $\overset{\beta}{\rightarrow}$ 4 Glu $\overset{\beta}{\rightarrow}$ Cer	Monosialohematoside (GM_3)
	NGNA 2 $\overset{\alpha}{\rightarrow}$ 3 Gal 1 $\overset{\beta}{\rightarrow}$ 4 Glu $\overset{\beta}{\rightarrow}$ Cer	
	NANA 2 $\overset{\alpha}{\rightarrow}$ 8 NANA 2 $\overset{\alpha}{\rightarrow}$ 3 Gal 1 $\overset{\beta}{\rightarrow}$ 4 Glu $\overset{\beta}{\rightarrow}$ Cer	Disialohematoside
	NGNA 2 $\overset{\alpha}{\rightarrow}$ 8 NGNA 2 $\overset{\alpha}{\rightarrow}$ 3 Gal 1 $\overset{\beta}{\rightarrow}$ 4 Glu $\overset{\beta}{\rightarrow}$ Cer	
	NGNA 2 $\overset{\alpha}{\rightarrow}$ 3 Gal 1 $\overset{\beta}{\rightarrow}$ 4 Glu $\overset{\beta}{\rightarrow}$ Cer ↑ AcO	
	NANA 2 $\overset{\alpha}{\rightarrow}$ 3 Gal 1 $\overset{\beta}{\rightarrow}$ 4 Glu $\overset{\beta}{\rightarrow}$ Cer ↑ AcO	

Table 2 (continued)

5. Ganglioside series	GalNAc $1 \xrightarrow{\beta} 4$ Gal $1 \xrightarrow{\beta} 4$ Glu $\rightarrow$ Cer $\uparrow \alpha$ NANA	GM_2; TAY-SACHS
	Gal $1 \xrightarrow{\beta} 3$ GalNAc $1 \xrightarrow{\beta} 4$ Gal $1 \xrightarrow{\beta} 4$ Glu $\rightarrow$ Cer $\uparrow 3$ $\uparrow \alpha$ NANA	GM_1
	Gal $1 \xrightarrow{\beta} 3$ GalNAc $1 \xrightarrow{\beta} 4$ Gal $1 \xrightarrow{\beta} 4$ Glu $\rightarrow$ Cer $\alpha \uparrow 3 \uparrow 3$ NANA$$NANA $\alpha \uparrow \uparrow 2$ (NANA)(NANA)	GD GT
	Gal $1 \xrightarrow{\beta} 4$ GluNAc $1 \xrightarrow{\beta} 3$ Gal $1 \xrightarrow{\beta} 4$ Glu $\rightarrow$ Cer $\alpha \uparrow $NGNA NGNA	Glucosamine Ganglioside (WIEGANDT)
6. Blood group glycolipid series	GluNAc $1 \xrightarrow{\beta} 3$ Gal $1 \xrightarrow{\beta} 4$ Glu $\rightarrow$ Cer	Wheat germ
	Gal $1 \xrightarrow{\beta} 4$ GluNAc $1 \xrightarrow{\beta} 3$ Gal $1 \xrightarrow{\beta} 4$ Glu $\rightarrow$ Cer	Type XIV
	Gal $1 \xrightarrow{\alpha} 3$ Gal $1 \xrightarrow{\beta} 3$ GluNAc $1 \xrightarrow{\beta} 3$ Gal $1 \xrightarrow{\beta} 4$ Glu $\rightarrow$ Cer	Rabbit "B"

Table 2 (continued)

$$\text{Gal } 1 \xrightarrow{\beta} 3 \text{ GluNAc } 1 \to 3 \text{ Gal } 1 \to 4 \text{ Glu} \to \text{Cer} \quad \text{Le}^a$$
$$\underset{\text{L Fuc}}{\overset{\alpha}{\uparrow}}$$

$$\text{Gal } 1 \xrightarrow{\beta} 4 \text{ GluNAc } 1 \to 3 \text{ Gal } 1 \to 4 \text{ Glu} \sim \text{Cer} \quad \text{"X-hapten"}$$
$$\underset{\text{L Fuc}}{\overset{\alpha}{\uparrow}}$$

$$\text{Gal } 1 \to 3 \text{ GluNAc } 1 \to 3 \text{ Gal } 1 \to 4 \text{ Glu} \to \text{Cer} \quad \text{Le}^b$$
$$\underset{\text{L Fuc}}{\overset{\alpha}{\uparrow}} \quad \underset{\text{L Fuc}}{\overset{\alpha}{\uparrow}}$$

Bloodgroup A, B, H glycolipid and higher glycolipid with long carbohydrate chain.

idea may be partially true; however, it should be remembered that erythrocytes are highly differentiated, metabolically less active cells and hence their glycolipid composition should be consistent. The glycolipid composition of cultured cells and other organs has shown great variety according to the conditions of culture and passage numbers [35], and it would not be surprising if it were affected by steroid hormones [51] and by the systemic metabolic imbalance which is observed in some organs of tumor-bearing animals [52].

BHK cells with lower passages showed better contact inhibition than those with higher passage numbers and contained a larger quantity of ceramide trihexoside [35]. Similarly 3T3 cells have a variety of glycolipid patterns and some 3T3 cells contain higher ganglioside [30] and some cells do not [29]. Variation of glycolipids in various cultured cells is shown in Table 3. Recent thorough studies by KLENK and CHOPPIN [55], RENKONEN et al. [56], DOD and GRAY [53] and WEINSTEIN et al. [54] have indicated that plasma membranes isolated from these cultured cells are greatly enriched in gangliosides and some neutral glycolipids. Recent work by YOGEESWARAN et al. [54a] has shown that gangliosides are synthesized in cells and located in plasma membrane.

The structural variations of glycolipids in animal cells are rather limited and may not exceed twenty or so; however, three-dimensional arrangements of glycolipids with other membrane components can offer almost unlimited possibilities of structural variation, so that there is a possibility that they could serve as an informational molecule on cell surfaces. A possible model of topographic arrangement is shown in Fig. 8. Two hydrophobic tails of glycolipids are associated with hydrophobic groupings of peptide chain and carbohydrate heads project outwards; some of the heads are masked by peptide chain, thus hampering their reactivities against enzymes or antibodies. The changes in the topography of glycolipids in membrane during ontogenetic development and by malignant transformation will be discussed later.

Glycolipids of Human Tumors

The heterologous antiserum directed against suspensions of human tumor cells reacts more strongly by complement fixation with the lipid component of human than with that of normal

Table 3. Glycolipids of various animal cells in culture

	BHK C13/21			BHK 21-W-12		BHK 21-F		3T3	3T3	
	(C1) W	(C2) W	(C3) W	PL	ER	PL	W	W	W	PL
CMH	±	±	±			±	+			
CDH	+		+			±				
CTH	+									
Globoside										
Forssman										
NAc-Hematoside	++	+	++	+++	±	+	++	+	++	++
NGly-Hematoside	+			++			++			
Disialohematoside	±									
Tay-Sachs (GM$_2$)										
Monosialoganglioside (GM$_1$)	±	±								
Disialoganglioside (GD)							+	+	+	++
Trisialoganglioside (GT)							±		+	+
References	[35]	[35]	[35]	[56]	[56]	[55]	[29]	[30]	[36]	[36]

Abbreviations: W = whole cells; PL = plasma membrane; ER = endoplasmic reticulum

Table 3 (continued)

	NIL-2K W	NIL-2E W	MDBK PL	HK PL	8166 W	SV-8161 W	CEF W	L-cell W	PL	L-929 PL
CMH	±									
CDH	± +	± +	+	+ +						
CTH	+ +	+ +	−	± + +						
Globoside	+ +	+ +	± +	+						
Forssman	+ +	± +	−	+						
NAc-Hematoside	+ +				+ +	±	+ +	+ +	+ +	+ +
NGly-Hematoside	+				−	−	−	−	−	+ +
Disialohematoside					+ +	+ +	+			+ +
Tay-Sachs (GM$_2$)		−	−	−	−					
Monosialoganglio- side (GM$_1$)	±		+	+	+ +	−	+	+	−	+
Disialoganglioside (GD)	−	−	+	+	−	−	+	+	± +	
Trisialoganglioside (GT)							±	−	−	±
References	[74]	[74]	[55, 74]	[55]	[35]	[35]	[78]	[54]	[54]	[54a]

tissue (WITEBSKY-HIRSZFELD phenomenon [57, 58]). The lipid haptens in the WITEBSKY-HIRSZFELD system were found by RAPPORT, in many cases, to be glycolipids, and that observation led to the isolation of cytolipin H, R and some other less well defined components [59, 60]. The chemical basis of this phenomenon is not known, as these glycolipids are usually found in normal tissue as well. The possibility that the phenomenon depends on organizational changes of glycolipid should be explored.

A glycolipid with a high content of fucose in addition to glucosamine, glucose and galactose was isolated from human adenocarcinoma [61, 62]. At that time this glycolipid was regarded as unique for human glandular tumor tissue, since sphingolipids with this carbohydrate composition had not been observed in human tissue before, although fucose and glucosamine are the most common components of glycoproteins. The glycolipid fraction showed inhibition for wheat germ agglutinin, H- and Le^a-hemagglutination [11]. Consequently, fucose-glycolipid of tumor tissue has been further fractionated on thin-layer chromatography as the acetylated derivative, as compared with that of normal glandular tissue. Among the glycolipids, two isomeric components have been isolated and their structure has been determined as shown in Fig. 1.

The following features of tumor glycolipid have become clear: absence of bloodgroups A and B, co-presence of Le^a and Le^b, and accumulation of Le^a-glycolipid and a glycolipid with unknown specificity (x-hapten). These have been observed irrespective of the bloodgroup status of the host [63, 64, 65].

The accumulation of disialohematoside and hematoside in human brain tumor has been described by SEIFERT and UHLENBRUCK [66], and UHLENBRUCK and GIELEN [67]. It is of great interest that these components are present in normal human brain at only a very low concentration. It is assumed, therefore, that the synthesis of carbohydrate chain in glycolipid of human tumor is blocked. As a consequence of the blockage, the accumulation of various precursor glycolipids could result; the accumulation could be an Le^a- or x-hapten glycolipid [64, 65], disialohematoside [66, 67], or TAY-SACHS' ganglioside (GM_2) [68], depending upon the sites of blockage occurring in the synthetic route of carbohydrate chain. In some human epithelial tumors, the accumulation of fucose-glycolipid did not occur and the blockage may not be obvious in glycolipid syn-

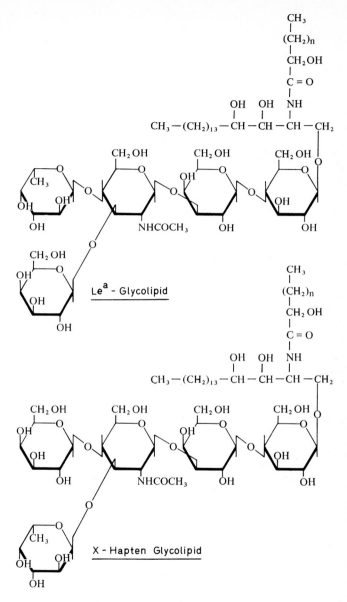

Fig. 1. Two isomeric glycolipid haptens accumulating in human tumor originated from glandular tissues

thesis [69], but no information is available as to whether any compensatory change could occur in glycoprotein. The structural alteration in glycoprotein of human epithelial tumor, as indicated by the specificity changes of bloodgroup, has been well established by several independent workers [5—10]. In studies of human tumors, however, difficulties, such as heterogeneity of tumor cells, and non-availability of comparable normal cells, have to be taken into consideration. The study has, therefore, been extended by the use of cultured cells and their *in vitro* transformants as will be described in the next section.

Glycolipids of Normal Cultured Cells and *in vitro* Transformed Cells

The first comparative experiments between normal "BHK", spontaneously transformed "BHK", and polyoma-transformed "BHK" cells has indicated that the quantity of hematoside was greatly reduced and that of lactosylceramide increased, namely the quantities of these two glycolipids are in a reciprocal relationship [28]. The degree of reduction of the hematoside quantity was greater in polyoma-transformed cells than those of spontaneously transformed cells. Both these results and those concerning bloodgroup change in human tumor tissue (see previous section) have supported a working hypothesis that the synthesis of carbohydrate chain in tumor cells in more often incomplete than their synthesis in normal cell surfaces ("incomplete synthesis model"). The accumulation of lactosylceramide was, however, not always as pronounced as reported in the first paper, this is partly due to a decrease of lactosylceramide level after successive passages [35].

In a subsequent study using BHK cells transformed by Rous sarcoma viruses and 3T3 cells transformed by SV-40 or SV-40 polyoma viruses, the level of hematosides decreased in these transformed cells. The level of glucosylceramide or lactosylceramide was unchanged or elevated [28, 29], an observation which again supported the view that synthesis of carbohydrate chain in the transformed cells is often incomplete (see Fig. 2).

A similar line of work was processed by BRADY and MORA and their coworkers, using 3T3 and AL/N mouse fibroblasts [30]. They observed that the quantity of di-sialo and monosialo-gangliosides (GM_1, GD_1) was reduced greatly in the transformed cells. They have claimed that the glycolipid pattern changed only in the presence of integrated viral genome and therefore spontaneously transformed cells without viral genome did not show any significant glycolipid changes [30]. This finding contrasts with our observation that the spontaneously transformed cells without viral genome still showed

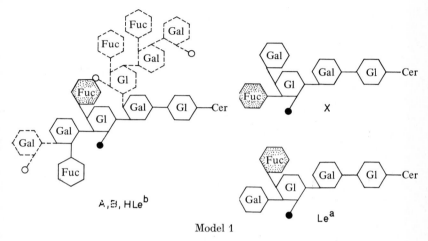

Fig. 2. Schematic expression of structural change of glycolipid: Structures expressed by dotted lines were deleted or reduced in their concentration, and the proximal structures (to ceramide side, which corresponds to the precursor) remained constant, or often their concentration increased. In many cases the part depicted as a dotted line completely disappeared, coupled with an accumulation of those structures described above. In human tumors of glandular tissues, accumulation of H, Le^a, or Le^b substance (shown in "Model 1") was probably coupled with the deletion of A and B haptens.

In experimental tumors, simpler patterns occur, as shown in "Model 2"

a significant glycolipid change. A decreased activity of UDP-N-acetylgalactosamine:hematoside N-acetylgalactosaminyl transferase was found in the homogenates of transformed cells [32]. The absence of disialoganglioside (GD_1a) in the SV-40 transformed fibroblasts has been confirmed by the work of SHEININ et al. [36]. They also observed that the plasma membrane of 3T3 and SV-3T3 contained all the ganglioside components including hematoside (GM_3), monosialoganglioside (GM_1) and disialoganglioside (GD_1a).

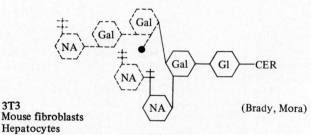

Fig. 2. Model 2

Glycoprotein and Total Heteroglycan Changes by Malignant Transformation

A systematic investigation was carried out by ROBBINS and his colleagues [37, 38] employing an ingenious double labelling experiment, i.e. normal 3T3 cells were labelled in the presence of ^{14}C-glucosamine and transformed cells were labelled by ^{3}H-glucosamine. The membrane heteroglycans were compared by gel filtration chromatography and by DEAE-cellulose chromatography. The results were rather complicated; however, they were interpreted as indicating the deletion of a non-reducing sialosyl group or N-acetyl-galactosaminosyl group in the heteroglycans of transformed 3T3 cells.

Recently, a similar technique has been used by BUCK et al. [39, 40] for the study of fucose-heteroglycans of BHK cells and of Rous sarcoma transformants. The cells were trypsinized and the material removed from the cell surface was digested by pronase. The material from the transformed cells contained an enrichment of glycopeptides of apparently higher molecular weight than the major group of glycopeptides from the control cells. This difference was also observed with the purified surface membranes isolated from control and transformed cells when analyzed by the same technique. An acrylamide gel electrophoresis of the solubilized plasma membrane fraction, prepared according to the WARREN procedure, showed a clear difference between the peptides of 3T3 cells and those of SV-3T3 cells [36]. The glycopeptides released by trypsin treatment from experimental tumor cells have been studied by LANGLEY and AMBROSE [70], WALBORG et al. [71] and by CODINGTON et al. [72]. The presence of a glucose-rich glycoprotein, reported by CODINGTON et al., would be of particular interest in view of a possible interaction with Con. A.

Glycolipid Changes in Slowly Growing, Non-glycolytic, Malignant Cells (Morris Hepatoma) as Compared With Rapidly Growing Normal Neonatal Liver

The incomplete synthesis described above could be simply a secondary phenomenon as a consequence of a rapid cell division rather than a primary essential step for the establishment of cancer, since the comparison has been made between slowly growing normal cells and rapidly growing transformed cells *in vitro*. In order to answer this question the glycolipid pattern of slowly growing Morris hepatoma has been compared to that of a rapidly growing normal fetal liver [33].

As shown in Table 4 and Fig. 3, trisialoganglioside (GT) has been completely deleted in either of three lines of Morris hepatoma (5123, 5123-C, 7800) and an enormous quantity of both disialoganglioside (GD_1a) and monosialoganglioside (GM_1) accumulated. It is noteworthy that the 5123 line is well known as showing low glycolysis

[33a]. Essentially the same type of change has been observed by MURRAY and his co-workers [34]; the deletion of trisialoganglioside (GT) and an increase of disialoganglioside in 7800 and 7777 minimal deviation hepatomas. The presence of trisialoganglioside (GT) seems

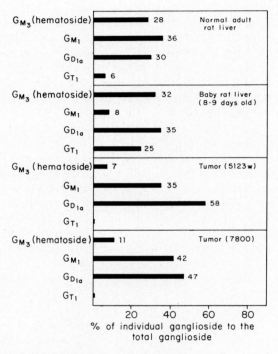

Fig. 3. Quantities of various gangliosides in minimal deviation hepatomas in comparison with normal liver and neonatal liver

to be characteristic of normal rat liver cells *in vivo*, since this glycolipid was absent in the cultured hepatocytes [31]. The glycolipids of cultured hepatoma cells deviated further from the glycolipid pattern of cultured hepatocytes. The hepatoma cells in culture were characterized by the absence of disialoganglioside (GD_1a) and an increased level of monosialoganglioside (GM_1). The trend of the changes of glycolipids in cultured hepatoma cells from those of

Table 4. Quantity of sialoglycolipids, neutral glycolipids and ceramide in normal livers and Morris hepatomas[a]

	Hematoside (GM$_3$)	Monosialo-ganglioside (GM$_1$)	Disialo-ganglioside (GD$_1$a)	Trisialo-ganglioside (GT)	Glucosyl-ceramide	Ceramide
	millimicromoles per milligram protein					
Normal adult rat liver	0.13	0.11	0.09	0.01	0.33	0.02
Baby rat liver	0.21	0.04	0.15	0.09		
Tumor 5123	0.64	2.57	3.60	0.00	0.93	0.43
Tumor 7800	0.57	1.87	1.73	0.00	0.83	0.41

[a] Reconstructed from SIDDIQUI and HAKOMORI [33]; all of these glycolipids and even ceramide were found in much higher concentration in plasma membrane (unpublished data of SIDDIQUI and HAKOMORI, see also ref. [53]).

hepatocytes was identical to that observed in tumors *in vivo* [33, 34]. The glycolipid changes observed in Morris hepatoma cells are of great significance in that the altered pattern of glycolipids may not be simply a result of rapid cell division or of the enhanced glycolytic activity[1] but the reflection of a primary change of an intrinsic and intricate property of the cell membrane.

Cell Density-dependent Glycolipid Changes: Glycosyl Extension Response and Lack of this Response in the Transformed Cells

The data described in the previous section have suggested that altered membrane glycolipid (or glycoprotein) may be linked to uncontrolled cell division. It should be noted, however, that the total amount of cellular heteroglycan was higher in cells grown on monolayer than in those grown in suspension culture, where cell-to-cell contact did not take place [73]. A related observation is that one of the membrane glycopeptides of mouse fibroblasts (3T3) increases upon cell confluence [38]. Consequently, the glycolipids of "normal" cultured cells with different degrees of contact inhibition were compared at different cell densities and with the glycolipids of virally transformed cells.

The α-galactosylglycolipids (CTH; see Table 2) of the contact-sensitive cells (cell line I; see Table 3) increased at high cell density and showed a higher labelling of CTH in cells at higher cell density [35]. In an other cell line which showed a lesser degree of contact inhibition and had no CTH (cell line II, III; see Table 3), the concentration of all the glycolipids increased at higher cell density (as judged by chemical analysis) [35]. Increased incorporation of isotope into the glycolipids (especially hematoside) at higher cell density was clearly demonstrated and this was more obvious than the results of chemical analysis. In contrast, the concentrations of glycolipids in two polyoma-transformed cells (I-py; III-py) did not change with cell density, as was shown by chemical analysis and by radiochemical analysis (for explanation, see Figs. 4 and 5).

---

[1] Neither 5123 nor 7800 cells are glycolytic [33a, 33b] and the production of lactic acid is at the normal level, therefore the change of glycolipid synthesis cannot be related to local pH changes due to the enhanced glycolytic activity. It is true, however, that many transformed cells *in vitro* have an enhanced glycolytic activity, and the culture media tends to become acidic much more quickly than with normal cells.

In human diploid fibroblasts, the concentration of monosialohematoside (GM_3) and disialohematoside and monosialoganglioside (GM_1) increased greatly with cell confluence. The glycolipids that

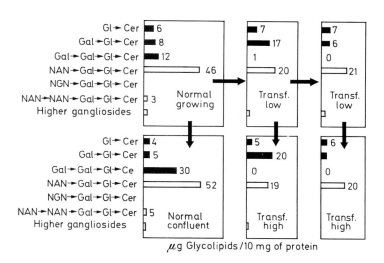

Fig. 4. Changes in glycolipid composition according to difference of growth phases and by transformation. This shows glycolipid changes of BHK C13/21-clone 1 cells which contain ceramide trihexoside. Note that the chemical quantity of this glycolipid greatly increased on cell confluence and was deleted by malignant transformation by polyoma virus. The second two rectangles shown the glycolipids of the py-transformed cells immediately after transformation. The third two rectangles show the glycolipids of the py-transformed cells after 50 passages. *Low* and *High* indicate the glycolipid quantities at low and high cell population densities. Note complete deletion of CTH and increase of CDH in freshly transformed cells. The increased quantity of CDH has disappeared after 50 passages, but the deletion of CTH and low amount of hematoside remain. Note also that there is no essential difference between low and high cell population densities

responded to cell confluence disappeared or were greatly reduced in concentration after SV-40 transformation. The SV-40 transformed cells had a ganglioside similar to GM_2 (TAY-SACHS's ganglioside). These results suggest that the concentration of certain glycolipids increases on cell-to-cell contact of normal cells. This is presumably due to the glycosyl extension reaction by addition of sugar residue to precursor lipid. The most typical and sensitive extension response was demonstrated by the terminal α-galactosyl residue of CTH in a contact-sensitive BHK cell line I and by the terminal sialosyl residue of disialosylhematoside of human diploid

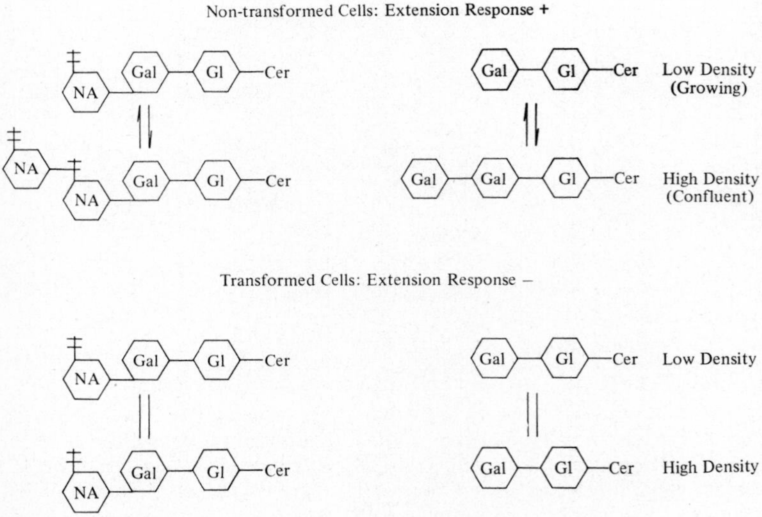

Fig. 5. Schematic expression of the presence of *"glycosyl extension response"* in normal cells and absence of it in transformed cells. This scheme shows how, in normal fibroblastic cells, hematoside or CDH could be converted on cell-to-cell contact to disialohematoside and CTH at high cell population density, while this ability is lacking in the transformed cells and the quantity of hematoside or CDH is constant, irrespective of the cell population density

cells [35]. The same glycolipid has been found in bovine kidney [35a] and in human gastrointestinal tract [68]. Robbins and Mac-Pherson have recently observed that much higher incorporation of ^{14}C-palmitate into CTH, and globoside, occurred when cultured cells were contact-inhibited and that incorporation became very low when these cells were in active growth [68a].

The activity of the biosynthetic enzyme for the terminal α-galactose residue of ceramide trihexoside (UDP-galactose:lactosylceramide α-galactosyltransferase) was greatly enhanced (2- to 3-fold increase) in contact-sensitive hamster BHK and NIL cells when the growth of these cells was contact-inhibited as compared to the same enzyme activity of the same cells at sparse cell density (Table 5). The activity of this enzyme in polyoma-transformed cells (either PY-BHK or PY-NIL-2E) was less than 50% of the activity of the

Table 5. *Activities of UDP-gal: lactosylceramide α-gal-transferase and UDP-gal:glucosylceramide β-galactosyltransferase of P-3 frac. prepared from normal and transformed cells*

		α-gal transferase μμmole/mg/hr	β-gal transferase μμmole/mg/hr
NIL-2E	sparse growing	385	246
	confluent	1218	252
PY-NIL	Low	26	642
	High	23	320

Cells were briefly sonicated with low intensity (60 watts for 90 seconds in "Biosonik"). The P-3 fraction was obtained from the supernatant of 12,000 xg centrifugation for 25 minutes and sedimented by 105,000 xg for 1 hour. The Assay system contained 0.05 μmole of either CMH or CDH, 800 μg of a detergent, "Cutscum", dissolved in 10 μl 1 M cacodylate pH 6.1. 20 μl 0.15 M manganese chloride, -^{14}C UDP-galactose (1.4 × 10^5 cpm/80 m μmoles, and 100 μl of the P-3 suspension). Details will be published elsewhere [74].

growing normal cells and was not influenced by cell population density [74].

In contrast, enzyme activity for lactosylceramide synthesis (UDP-gal:glucosylceramide β-galactosyltransferase) was not overly dependent on cell population density and was less affected by malignant transformation. The activity of a degradative enzyme for the terminal α-galactose residue of ceramide trihexoside was not affected by cell population density, but was enhanced in the transformed cells [74].

The results indicate that α-galactosyl synthesis could be enhanced on cell-to-cell contact. Interestingly, the glycolipid that shows cell density-dependent positive responses decreases in concentration on successive passages or after malignant transformation of cells. The "incomplete" synthesis of the carbohydrate chain apparently observed in the transformed cells may be related to a lack of "glycosyl extension response", i.e. lack of addition of terminal sugar residue in the transformed cells on cell-to-cell contact.

Time Course of Glycolipid Changes During Transformation After Infection with Rous Sarcoma Virus

Cultures of chick embryonic fibroblasts and transformation by Rous sarcoma virus [75, 76] have been used for this comparison. A great advantage of this system is that over 90% of the cells can be transformed within 1 to 2 days [77]. Rous sarcoma virus-induced changes of glycolipid can then be studied before prolonged periods of cell culture superimpose additional alterations which may be unrelated to virus-induced neoplastic transformation. This system also made it possible to use the same chick embryo cells for infected and for control cultures; variations due to genetic heterogeneity of cells can therefore be eliminated. Fig. 6 shows the time course change of hematoside, disialohematoside and an unidentified sialosyl lipid. A remarkable progressive decrease of SL-2 and SL-3 has been noticed during the process of transformation. Quantitative data of the glycolipid changes are presented in Fig. 6. It is interesting to note that the decrease of disialohematoside was more sensitive and started earlier after the infection than monosialohematoside. The decrease of monosialohematoside occurred obviously after 60 h of infection and was obviously at a lower level than that of normal controls when the transformation was completed [78]. Fig. 7 also shows the change of ceramide and glucosylceramide; these lipids stayed constant during transformation but showed an increased level when the transformation was completed [78]. It is also noteworthy that the chemical quantity of hematoside and disialohematoside increased at higher cell population density, whereas those glycolipids that show cell-density-dependent changes decrease in concentration during malignant transformation. Thus, a similar phenomenon as was described in the previous section has again been demonstrated.

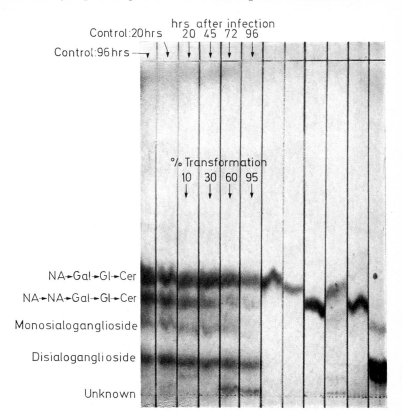

Fig. 6. Time-dependent changes of glycolipid in chick fibroblasts after infection with Rous sarcoma virus. Note progressive decrease of hematoside, disialohematoside, and an uncharacterized glycolipid corresponding to monosialoganglioside, 20, 45, 72, and 96 hours after infection with Rous sarcoma virus. The approximate transformation rates were 10, 30, 60, and 95% of the population, respectively. Note also an increase of hematoside and disialohematoside after the cell sheets were left for 90 hours

Reactivities of Cell Surface Glycolipids to Their Antisera

The quantity of glycolipids chemically determined in cell membrane is not related to the reactivity of cells to the antiserum directed against the glycolipid; e.g. globoside (see Table 2) is the major component of human erythrocyte membrane and constitutes more than 80% of the total glycolipid of human erythrocyte membrane, but anti-globoside antiserum does not react with human erythrocytes [79]. In contrast, bloodgroup A and B glycolipids, which are

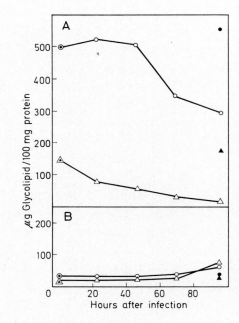

Fig. 7. Quantitative changes of monosialohematoside (0-0-0), disialohematoside (Δ-Δ-Δ) in A and glucosylceramide and ceramide in B during the course of complete transformation by Rous sarcoma viruses

the minor components of human erythrocyte membrane are strongly antigenic and highly reactive to anti-A or anti-B antisera.

The non-reactive globoside of human erythrocytes becomes highly reactive when human erythrocytes are treated with trypsin or pronase [79, 80], and perhaps the globoside group is in cryptic state in normal erythrocyte membrane, being covered by protein layer(s). A similar result has been found by UHLENBRUCK et al. using invertebrate agglutinin which is directed to N-acetylgalactos-

Table 6. *Reactivities of fetal and adult erythrocytes to anti-globoside antiserum*[a]

	Fetal (mean of 7 cases)	Adult (mean of 10 cases)
Agglutination titer		
native	220	10
trypsinized	500	320
Hemolysis titer		
native	320	20
trypsinized	640	320
Absorption		
native	10.5 μgN[b]	1.4 μgN[b]

[a] Abstracted from HAKOMORI [ref. 80].
[b] The amount of anti-globoside antibody absorbed by 1 ml of packed erythrocytes.

Table 7. *Reactivities of anti-hematoside antiserum to "normal" and "transformed" cells*[a]

Reactivity
Degree of Cr^{51} release: % cpm
(Released cpm/total cell-bound cpm)

	Native	Trypsinized	Δ%
BHK	42	75	45% increase
3T3	49	70	30% increase
PYBHK	70	70	0
SV 3T3	69	65	0

[a] Abstracted from HAKOMORI, TEATHER and ANDREWS [19].

amine residue [81]. Erythrocytes of human fetus (2 to 4 months old) obtained from early abortion cases was found to be highly reactive to the anti-globoside antiserum, in striking contrast to newborn cord erythrocytes and adult erythrocytes, although globoside levels of fetal erythrocytes are in the same range as in newborn cord or adult erythrocytes [80] (see Table 6). The results strongly suggest that, during the process of ontogenetic development of erythrocytes, the globosidic groups have become cryptic. The erythrocyte does not show any serological activity of globosides after neonatal life.

Consequently, in view of the similarity of oncogenic and ontogenetic processes, the reactivity of normal and transformed BHK and 3T3 cells to anti-hematoside serum has been investigated, since hematoside is the principal glycolipid of these cultured fibroblastic cells.

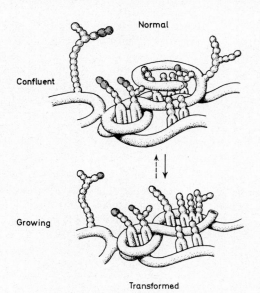

Fig. 8. Idealized version of the organization of glycolipids with peptide chains on the surface of plasma membrane. Different kinds of glycolipids could be arranged in order, being bound with hydrophobic bonds with protein. The hydrophilic carbohydrates could face outwards, some of them being covered by peptide(s). Some glycolipids contain two repeating carbohydrates with α-linkage; these can change their structure more easily than do other structures on cell-to-cell contact, as well as on transformation. The essential changes on transformation are shown in this figure as removal of the peptide cover and also removal of the terminal sugar, especially those of repeating structures

The reactivity of cells to anti-hematoside antiserum was determined by release of Cr^{51} upon immunolysis of the labelled cells. As is shown in Table 7, normal 3T3 cells or BHK cells had very low activity and showed about 30—50% increase of activity after trypsin treatment [29]. The transformed 3T3 or BHK cells, in contrast to non-transformed cells, showed a definite reactivity and the activity was not enhanced by proteolysis [29].

The uncovering of cryptic carbohydrates on cell surfaces which occurs during the process of oncogeny has been extensively elucidated by BURGER [82] using wheat germ phytoagglutinin, and later by INBAR and SACHS [83] using Concanavalin A. HÄYRY and DEFENDI's [84] studies using anti-3T3 cell antiserum have also supported the unmasking of cryptic carbohydrates, although the reactive sites for anti-transformed 3T3 cells are not related to either hematosidic or wheat germ agglutinin reactive sites.

An extremely exciting view has been further developed by BURGER who showed that a contact-inhibited monolayer of normal 3T3 cells shows further proliferation in the presence of a low concentration (0.007%) of trypsin [85] and that the growth of polyoma-transformed 3T3 cells was normalized by coating with trypsin-treated concanavalin A [86]. The unmasking of the reactive sites of cells to phytoagglutinin seems to occur during mitotic phase and is masked on stationary phase [87], thus suggesting a link between masking and/or unmasking of carbohydrate sites on cell surfaces and nuclear DNA replication.

Epilogue

Two major cell surface changes associated with malignant transformation have become obvious: 1) the incomplete synthesis of carbohydrate chain which is due to a lack of glycosyl extension response on cell-to-cell contact, and 2) the consistent uncovering of some carbohydrate sites on tumor cell surfaces which is probably due to a lack of transitional changes between covering and uncovering status, according to the cell cycle.

Both phenomena indicate that they are possibly linked with the regulation of nuclear DNA replication, i.e. initiation or termination of DNA synthesis. The behavioral normalization of transformed cells by binding with trypsinized concanavalin A, a startling discovery by BURGER, has pinpointed that the malignancy of cells is a regulatory dysfunction of the surface membrane. It may not be necessary to further discuss BURGER-SACHS dogma, but would be sufficient to mention the significance of incomplete carbohydrate chains, especially those of α-glycosyl linkages.

The composition of carbohydrate chains, e.g. those containing α-galactosylgalactose and the sialosylsialosyl group, should be quite variable according to the cell population density and the status of cell growth; they can extend their chain length when cells are in contact with each other through activation of sugar nucleotide transferase (e.g. UDP-galactose α-galactosyltransferase), and they may reduce their chain length when cells are in a growing phase (or, more exactly, when they are in a dividing phase).

Contact-sensitive carbohydrate chains, especially those containing an α-glycosidic linkage, may serve to bind cell to cell, as has been suggested by ROSEMAN [88], or, on the other hand, to change the conformation of a "repeating unit" of cell membrane, hence to change the "cooperativity" of a membrane subunit which may eventually affect nuclear DNA replication.

Transformed cells obviously lack both such intercellular adhesion and also the surface control mechanism for cell division, therefore they very often lack contact-sensitive linkages, such as α-galactosylgalactosyl, sialosylgalactosyl or sialosylsialosyl. The incomplete synthesis of carbohydrate chains observed in transformed cells is thus accounted for. Since not all transformed cells, however, show changes in their glycolipids, it is assumed that some cells may have contact-sensitive groups in their glycoproteins. Such cells may change their glycoprotein rather than change their glycolipids during malignant transformation.

Incompleteness of carbohydrate chains is sometimes accompanied by an accumulation of the precursor glycolipid, or a glycoprotein having a precursor chain; thus, sometimes a unique carbohydrate chain, which is not present or present only in trace amounts in non-transformed cells, can be recognized. Such an accumulation of a unique carbohydrate chain may be related to the change in immunological specificities of tumor cells.

The incomplete carbohydrates are not simply a result of rapid cell growth nor of lowered local pH due to the increasing glycolysis of malignant cells, as the non-glycolytic Morris hepatoma showed a "malignant pattern" in contrast to that of neonatal and adult liver. The changes of glycolipid started in the early stages during the transformation process, as was shown in the process of the transformation by Rous sarcoma virus.

Thus, as shown in Fig. 8, the mechanisms of carbohydrate chain extension and the covering/uncovering of carbohydrate sites cooperate during cell growth and control cell division. Thus, malignant cells are probably characterized by the lack of either or both of these mechanisms, which leads to the loss of membrane-mediated control of cell divisions.

Acknowledgement

The author is gratefully indepted to the following persons who have been cooperating in these projects: Dr. H. YANG (presently at National Research Council, Canada), Dr. T. SAITO (presently at Tohoku University, Sendai), Dr. H. KAWAUCHI (presently at Tohoku College of Pharmacy, Sendai), Dr. BADER SIDDIQUI, Dr. SHIGEKO KIJIMOTO, Mr. HENRY ANDREWS, Mrs. CAROL TEATHER and Mr. PHILLIP LU.

The work described under "Time course changes of glycolipid" was carried out in cooperation with Professor P. K. VOGT, Department of Microbiology, University of Washington.

The author is most grateful to the Deutsche Biochemische Gesellschaft and the organizers of this colloquium for providing the opportunity at Mosbach.

References

1. SACHS, L.: Nature (Lond.) **207**, 1272 (1965).
2. ABERCROMBIE, M.: Europ. J. Cancer **6**, 7 (1970).
3. EDWARDS, J. G., CAMPBELL, J. A.: J. Cell Sci. **8**, 666 (1970).
4. LOEWENSTEIN, W. R.: Develop. Biol. Suppl. **2**, 151 (1963).
5. MASAMUNE, H., KAWASAKI, H., ABE, S., OYAMA, K., YAMAGUCHI, Y.: Tohoku J. exp. Med. **68**, 81 (1958).
6. — HAKAMORI, S.: Symposia of Cell Chemistry **10**, 37 (1960).
7. KAY, H. E. H., WALLACE, B. M.: J. nat. Cancer Inst. **26**, 1349 (1966).
8. DAVIDSOHN, I., KOVARIK, S., LEE, C. L.: Arch. Path. **81**, 381 (1966).
9. KAWASAKI, H.: Tohoku J. exp. Med. **68**, 119 (1958).
10. ISEKI, S., FURUKAWA, K., ISHIKAWA, K.: Proc. Jap. Acad. **38**, 556 (1962).
11. HAKOMORI, S., KOSCIELAK, J., BLOCH, K. J., JEANLOZ, R. W.: J. Immunol. **98**, 31 (1967).
12. GOLD, P., FREEDMAN, S. W.: J. exp. Med. **121**, 439; **122**, 467 (1965).
13. — GOLD, M., FREEDMAN, S. W.: Cancer Res. **28**, 1731 (1968).
14. KLEIST, S. V., BURTIN, P.: Immunology **10**, 507 (1966).
15. KARITZKY, D., BURTIN, P.: Europ. J. Biochem. **1**, 411 (1967).
16. STONEHILL, E. H., BENDICH, A.: Nature (Lond.) **228**, 370 (1970).
17. HÄKKINEN, I.: Zut. J. Cancer **3**, 582 (1968).
18. — J. nat. Cancer Inst. **44**, 1183 (1970).
19. HAKOMORI, S., ANDREWS, H. D.: Biochim. biophys. Acta (Amst.) **202**, 1225 (1970).
20. KLEIN, G.: Israel J. med. Sci. **7**, 111 (1971).
21. FOGEL, M., SACHS, L.: J. nat. Cancer Inst. **29**, 239 (1962).

22. O'NEILL, C. H.: J. Cell Sci. **3**, 405 (1968).
23. ROBERTSON, H. T., BLACK, P. H.: Proc. Soc. exp. Biol. (N. Y.) **130**, 363 (1969).
24. FORRESTER, J. A., AMBROSE, E. J., MACPHERSON, I. A.: Nature (Lond.) **196**, 1068 (1962).
25. RUHENSTROTH-BAUER, G., KÜBLER, W., FUHRMANN, G. F., RUEFF, F.: Klin. Wschr. **39**, 764 (1964).
26. DEFENDI, V., GASIC, G.: J. cell. comp. Physiol. **62**, 23 (1963).
27. OHTA, N., PARDEE, A. B., MCAUSLAN, B. R., BURGER, M. M.: Biochim. biophys. Acta (Amst.) **158**, 98 (1968).
28. HAKOMORI, S., MURAKAMI, W. T.: Proc. nat. Acad. Sci. (Wash.) **59**, 254 (1968).
29. — TEATHER, C., ANDREWS, H. D.: Biochem. biophys. Res. Commun. **33**, 563 (1968).
30. MORA, P. T., BRADY, R. O., BRADLEY, R. M., MCFARLAND, V. W.: Proc. nat. Acad. Sci. (Wash.) **63**, 1290 (1969).
31. BRADY, R. O., BOREK, C., BRADLEY, R. M.: J. biol. Chem. **244**, 6552 (1969).
32. CUMAR, F. A., BRADY, R. O., KOLODNY, E. H., MACFARLAND, V. W., MORA, P. T.: Proc. nat. Acad. Sci. (Wash.) **67**, 757 (1970).
33. SIDDIQUI, B., HAKOMORI, S.: Cancer Res. **30**, 2930 (1970).
33a. AISENBERG, A. C., MORRIS, H. P.: Nature (Lond.) **191**, 1314 (1961).
33b. WEINHOUSE, S.: In: Biological and Biochemical Evaluation of Malignancy in Experimental Hepatomas. Gann Monograph Nr. 1, pp. 99 to 115. Edited and published by The Japanese Foundation for Cancer Research, Tokyo, 1966.
34. CHEEMA, P., YOGEESWARAN, G., MORRIS, P. M., MURRAY, R. K.: FEBS Letters **11**, 181 (1970).
35. HAKOMORI, S.: Proc. nat. Acad. Sci. (Wash.) **67**, 1741 (1970).
35a. PURO, K.: Biochim. biophys. Acta (Amst.) **189**, 401 (1969).
36. SHEININ, R., ONADERA, K., YOGEESWARAN, G., MURRAY, R. K.: Second Le Petit Symposium — The Biology of Oncogenic Viruses (in press).
37. WU, H., MEEZAN, E., BLACK, P. H., ROBBINS, P. W.: Biochemistry **8**, 2509 (1969).
38. MEEZAN, E., WU, H. C., BLACK, P. H., ROBBINS, P. W.: Biochemistry **8**, 2518 (1969).
39. BUCK, C. A., GLICK, M. C., WARREN, L.: Biochemistry **9**, 4567 (1970).
40. — — — Science **172**, 169 (1971).
41. BURGER, M. M.: Proc. nat. Acad. Sci. (Wash.) **62**, 994 (1969).
42. INBAR, M., SACHS, L.: Nature (Lond.) **223**, 710 (1969).
43. POLLACK, R. E., BURGER, M. M.: Proc. nat. Acad. Sci. (Wash.) **62**, 1074 (1969).
44. HÄYRY, P., DEFENDI, V.: Virology **41**, 22 (1970).
45. SELA, B.-A., LIS, H., SHARON, N., SACHS, L.: J. Membrane Biol. **3**, 267 (1970).
46. CARTER, H. E., JOHNSON, P., WEBER, E. J.: Ann. Rev. Biochem. **34**, 109 (1965).

47. WIEGANDT, H.: Angew. Chemie Internat., Ed., **7**, 87 (1968).
48. YAMAKAWA, T.: In: Lipoide 16 Colloquium der Gesellschaft Physiol. Chemie, Mosbach/Baden, 1965, p. 87 (SCHUTTE, E., Ed.). Berlin-Heidelberg-New York: Springer 1966.
49. HAKOMORI, S., SIDDIQUI, B., LI, Y-T., LI, S.-C., HELLERQVIST, C. G.: J. biol. Chem. **246**, 2271 (1971).
50. SIDDIQUI, B., HAKOMORI, S.: J. biol. Chem. **246**, 5766 (1971).
51. COLES, L., HAY, J. B., GRAY, G. M.: J. Lipid Res. **11**, 158 (1970).
52. ADAMS, E. P., GRAY, G. M.: Nature (Lond.) **216**, 278 (1967).
53. DOD, B. J., GRAY, G. M.: Biochim. biophys. Acta (Amst.) **150**, 397 (1968).
54. WEINSTEIN, D. B., MARSH, J. B., GLICK, M. C., WARREN, L.: J. biol. Chem. **245**, 3928 (1970).
54a. YOGEESWARAN, G., WHERRETT, J. R., CHATTERJEE, S., MURRAY, R. K.: J. biol. Chem. **245**, 6718 (1971).
55. KLENK, H. D., CHOPPIN, P. W.: Proc. nat. Acad. Sci. (Wash.) **66**, 57 (1970).
56. RENNKONEN, O., GAHMBERG, C. G., SIMONS, K., KÄÄRIÄINEN, L.: Acta chem. scand. **24**, 733 (1970).
57. WITEBSKY, E.: Z. Immunitätsforsch. **62**, 35 (1929).
58. HIRSZFELD, L., HALBER, W., LASKOWSKI, Z.: Z. Immunitätsforsch. **63**, 81 (1929).
59. RAPPORT, M. M., GRAF, L.: Cancer Res. **21**, 1225 (1961).
60. — — Prog. Allergy **13**, 273 (1969).
61. HAKOMORI, S., JEANLOZ, R. W.: J. biol. Chem. **239**, pc 3606 (1964).
62. — Blood and tissue antigen, p. 149 (AMINOFF, D., Ed.). New York: Academic Press 1970.
63. — Chem. Phys. Lipid **5**, 96 (1970).
64. — ANDREWS, H. D.: Biochim. biophys. Acta (Amst.) **78**, 313 (1963).
65. YANG, H., HAKOMORI, S.: J. biol. Chem. **246**, 1192 (1971).
66. SEIFERT, H., UHLENBRUCK, G.: Naturwissenschaften **32**, 190 (1965).
67. UHLENBRUCK, G., GIELEN, W.: Med. Welt **20**, 332 (1969).
68. KAWAUCHI, H., HAKOMORI, S.: In preparation.
68a. ROBBINS, P. W., MACPHERSON, I.: Nature (Lond.) **229**, 669 (1971).
69. KAWANAMI, J., TSUJI, T.: Jap. J. exp. Med. **38**, 11 (1968).
70. LANGLEY, O. K., AMBROSE, E. J.: Nature (Lond.) **204**, 53 (1964).
71. WALBORG, E. F., LANTZ, R. S., WRAY, V. P.: Cancer Res. **29**, 2034 (1969).
72. CODINGTON, J. F., SANFORD, B. H., JEANLOZ, R. W.: J. Nat. Cancer Institute **45**, 637 (1970).
73. SHEN, L., GINSBRUG, V.: In: Biological properties of mammalian surface membrane (MASON, L. A., Ed.). Wistar Institute Symposium Monograph Nr. 8, p. 67 (1968).
74. KIJIMOTO, S., HAKOMORI, S.: Biochem. biophys. Res. Commun. **44**, 557 (1971).
75. RUBIN, H.: Proc. nat. Acad. Sci. (Wash.) **46**, 1105 (1960).
76. VOGT, P. K., RUBIN, H.: Virology **19**, 92 (1963).
77. HANAFUSA, H.: Proc. nat. Acad. Sci. (Wash.) **63**, 313 (1969).
78. HAKOMORI, S., SAITO, T., VOGT, P. K.: Virology (in press).

79. KOSCIELAK, J., HAKOMORI, S., JEANLOZ, R. W.: Immunochemistry 5, 441 (1968).
80. HAKOMORI, S.: Vox Sang. (Basel) 16, 478 (1969).
81. UHLENBRUCK, G., REIFENBERG, V., HEGGEN, M.: Z. Immun- und Allergie-Forsch. 139, 486 (1970).
82. BURGER, M. M.: Proc. nat. Acad. Sci. (Wash.) 62, 994 (1969).
83. INBAR, M., SACHS, L.: Nature (Lond.) 223, 710 (1969).
84. HÄYRY, P., DEFENDI, V.: Virology 41, 22 (1970).
85. BURGER, M. M.: Nature (Lond.) 227, 170 (1970).
86. — NOONAN, K. D.: Nature (Lond.) 228, 512 (1970).
87. FOX, T. O., SHEPPARD, J. R., BURGER, M. M.: Proc. nat. Acad. Sci. (Wash.) 68, 244 (1971).
88. ROSEMAN, S.: Chem. and Phys. Lipids 5, 270 (1970).

Structure and Biosynthesis of Viral Membranes

HANS-DIETER KLENK

*Institut für Virologie der Medizinischen Fakultät der Justus Liebig-Universität
D-6300 Giessen*

With 9 Figures

Many animal viruses are coated by an envelope which is an integral component of the virion. During virus maturation this membrane envelops the internal component, the nucleocapsid. Upon degradation of the envelope, these viruses have lost infectivity and other biological functions. Morphologically the envelopes resemble cellular membranes, on electron micrographs they appear as unit membranes. The envelope may be formed in the cytoplasm without connexion to cellular membranes, as in the case of vaccinia virus (DALES and MOSBACH, 1968). Most viruses, however, are assembled in close continuity with cellular membranes and are then released in a budding process. For instance, the envelope of herpes virus is formed at the nuclear membrane (SIEGERT and FALKE, 1966), the envelope of certain rhabdoviruses at the membranes of intracellular vacuoles (MUSSGAY and WEIBEL, 1963), and the envelopes of arboviruses (ACHESON and TAMM, 1967), RNA-tumor viruses (BEARD et al., 1963) and myxoviruses at the plasma membrane of the host cell.

Analysis of the composition of membrane-enclosed viruses and of the morphological and biochemical events in virus assembly is obviously necessary for an understanding of viral structure and replication. Furthermore, this information may shed some light on cell membrane structure and biogenesis in general. This report describes studies dealing with influenza viruses and the parainfluenza virus SV5.

Morphological Aspects of Myxovirus Assembly

Detailed investigations of the morphology and replication of SV5 have been carried out in recent years by P. W. Choppin and his group at the Rockefeller University in New York (Choppin and Stockenius, 1964; Compans et al., 1966). These findings, which are also valid to a large extent for influenza virus, will be briefly reviewed here.

SV5 virions consist of a lipoprotein envelope covered by a layer of surface projections or spikes, which are typical of all myxo- and paramyxoviruses. The viral envelope is antigenically strain-specific and it contains hemagglutinin, neuraminidase, and a factor which induces cell fusion and hemolysis. Coiled within the envelope is the nucleocapsid, a single-helical ribonucleoprotein. It has a unit length of 1 μ, and a diameter of 180 Å (Compans and Choppin, 1967). It is composed of a single polypeptide species (Caliguiri et al., 1969; Mountcastle et al., 1970), and contains single-stranded RNA with a molecular weight of 6 to 7 million (Compans and Choppin, 1968). Highly purified SV5 virions are composed of 0.9% RNA, 73% protein, 20% lipid, and 6% carbohydrate [Klenk and Choppin. 1969 (1)].

The basic principles of assembly of SV5 at the cell membrane have been described in detail (Compans et al., 1966). The following discussion is based on this earlier work and recent studies with SV5 and influenza virus (Choppin et al., 1971; Compans and Dimmock,

---

Fig. 1. Monkey kidney cells 22 h after inoculation with SV5. a) Strands of nucleocapsid free in the cytoplasmic matrix near the surface of the cell (arrows). Magnification: × 47500. b) Nucleocapsid strands, many in cross section, closely aligned under a long region of the cell membrane. A layer of dense material resembling the spikes found in the viral envelope is present on the outer surface of the membrane (arrow). Magnification: × 73000. c) Two particles budding at the cell surfaces which contain nucleocapsid and possess spikes on the outer surface of the unit membrane (arrows). Magnification: × 52500. d) A row of eight budding particles showing many cross sections of nucleocapsid. Magnification: × 63000. e) and f) Long filaments protruding to the cell surface. Viral surface projections and nucleocapsid strands are present along the entire length of filaments but are absent on the adjacent cell membrane; the layers of the cell membrane are continuous with those of the unit membrane in the viral envelope (arrows). Magnification: e) × 94000; f) × 131000. (From Compans et al., 1966)

Structure and Biosynthesis of Viral Membranes 99

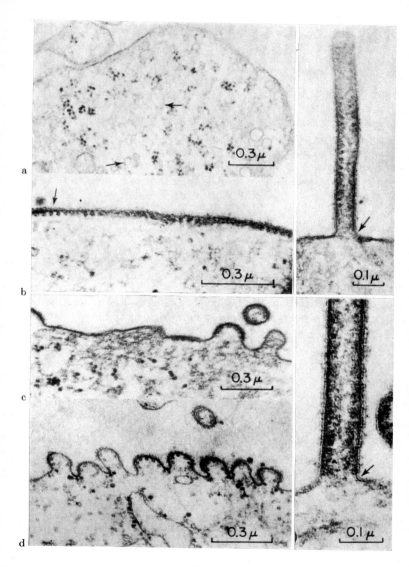

Fig. 1a—f

1969) concerning the events at the cell membrane. The first morphologically identifiable step in virus assembly is the appearance of the nucleocapsid in the cytoplasm (Fig. 1). The nucleocapsid then becomes aligned in a regular array beneath areas of cell membrane which have a layer of spikes on their outer surface. Neither the aligned nucleocapsid nor the spikes are seen alone.

The next step in virus assembly is the well-known budding process. Both roughly spherical virions and long virus filaments are produced by this process. An important feature of the assembly of the virus is that during the budding process the membrane of the viral envelope is continuous with and morphologically similar to the plasma membrane of the host cell. Not only does the membrane of SV5 virions resemble that of the cell, but the membranes of virions grown in different cells, whose membranes show distinctive staining properties, can also be distinguished.

Polypeptides of Parainfluenza Virus SV5

It has been known for some time that myxoviruses contain three biologically active proteins: one protein which forms a complex with the viral RNA and is responsible for the group-specific antigenicity, and two surface proteins which have hemagglutinin and neuraminidase properties. With the advent of polyacrylamide gel electrophoresis, a method for separation of polypeptides on the basis of their size, it became evident that these viruses contained more than three peptides.

Polyacrylamide gel electrophoresis of purified SV5 virions dissociated with sodium dodecyl sulfate and mercaptoethanol (CALIGUIRI et al., 1969) revealed six polypeptides (Fig. 2), the largest of which appears to be present in very small amounts so that it was not always detected in radioactively labeled gels; however, this protein was consistently found in stained gels of virus grown in four different cells. Double label experiments with two amino acids were done to analyze the ratios across each of the six peaks in the gels in an attempt to detect other polypeptides beneath these peaks and to investigate the variable amount of material at the origin of some gels. A constant ratio of the two isotopes was found across the various peaks, suggesting that each represented a single polypeptide, and the variable ratio across the material at the origin confirmed that this represented aggregation.

The molecular weights of the virion polypeptides have been estimated by coelectrophoresis with marker proteins and range from 43000 to 76000. The estimated size of the single-stranded RNA genome of SV5 is about 6 million. There is thus more than enough genetic material to code for the 350000 daltons of all six polypeptides found in the SV5 virion.

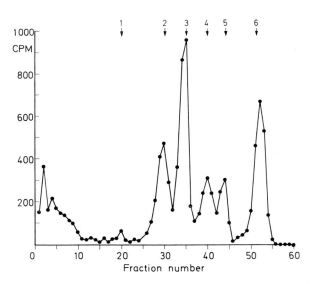

Fig. 2. Polyacrylamide gel electrophoresis of SV5 virion polypeptides labeled with ^{3}H-amino acid mixture. The virus was grown in MK cells. Six polypeptides, one a very minor component, are present. (From CALIGUIRI et al., 1969)

This pattern of polypeptides has been found consistently, not only in virus grown in monkey kidney (MK) cells, but also in virus grown in three other cell types: bovine kidney cells (MDBK), and two lines of hamster kidney cells (BHK 21-F and HaK). Furthermore, on coelectrophoresis of isotopically labeled SV5 virions grown in MK and MDBK cells, the proteins of the virions grown in the two cells migrated identically in each case (CHOPPIN et al., 1971). This indicates that the polypeptides of virions grown in different host cells have the same size. These findings and the results from experiments in which virus was grown in prelabeled cells, suggest that the structural proteins of the virion are coded by the viral genome. Although the virus makes use of the plasma membrane of the host cell during its assembly, there appears to be no host cell protein in the virion. Similar conclusions have been reached for other viruses, in the envelopes of which also exclusively virus specific proteins could be detected (HOLLAND and KIEHN, 1970; COMPANS et al., 1970).

Carbohydrates of SV5

As a further step in the characterization of these polypeptides, an examination of their carbohydrates content was undertaken (KLENK et al., 1970). SV5 contains about equal amounts of hexose and hexosamine and smaller amounts of fucose. Two thirds of the carbohydrate is covalently bound to protein, and one third is bound to lipid. Similar carbohydrate compositions have been reported for influenza (FROMMHAGEN et al., 1959; ADA and GOTTSCHALK, 1956) and Sindbis virus (STRAUSS et al., 1970). There is also evidence for the presence of carbohydrates in vesicular stomatitis virus (VSV) (MC SHARRY and WAGNER, 1971; BURGE and HUANG, 1970), herpesviruses (OLSHEVSKY and BECKER, 1970; SPEAR et al., 1970; BEN-PORAT and KAPLAN, 1970), RNA tumor viruses (BOLOGNESI and BAUER, 1970), and two other myxoviruses (MOUNTCASTLE et al., 1971). However, those viruses which do not belong to the myxovirus group have been found to contain significant amounts of neuraminic acid. SV5, which like all myxoviruses contains the enzyme neuraminidase, has no neuraminic acid, either protein-bound or lipid-bound.

In order to get more detailed information on the carbohydrate composition of the virus and on the carbohydrate content of the various structural proteins of the virion, SV5 was labeled with radioactive amino acids and a variety of radioactive sugars. The virus was then purified and (1) subjected to acid hydrolysis followed by thin-layer chromatography to identify the constituent sugars, and (2) dissociated with SDS-mercaptoethanol and the proteins were examined by polyacrylamide gel electrophoresis [KLENK et al., 1970 (1)]. The studies revealed that the virus contains galactose, mannose, glucosamine, and fucose linked to viral proteins. Glucose, galactosamine, and galactose are constitutents of glycolipids in the virion, as discussed below.

Polyacrylamide gel electrophoresis of virions grown in the presence of radioactive sugars revealed that two of the envelope proteins of SV5 are glycoproteins. ^{3}H-Glucosamine was incorporated exclusively into polypeptides 2 and 4 (Fig. 3). There was no labeling of the other proteins. Similar results were obtained when ^{3}H-fucose

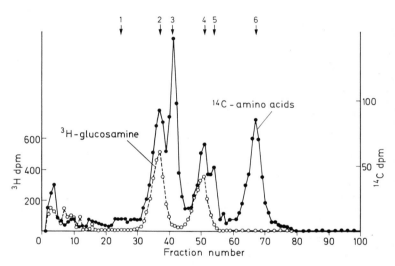

Fig. 3. Polyacrylamide gel electrophoresis of SV5 polypeptides labeled with ^{3}H-glucosamine and ^{14}C-amino acids. Virion polypeptides 2 and 4 contain the carbohydrate label. [From KLENK et al., 1970 (1)]

was used as the labeled monosaccharide. These sugars have been shown to be specific markers in this system, i.e. they are incorporated into the virion as such, with little or no breakdown and re-utilization of the label in other substances [KLENK et al., 1970 (1)].

One of the most obvious questions raised by the existence of glycoproteins in the virus particle is that of the specificity of the carbohydrate portion; is it specified by the host, the viral genome, or some combination of both ? At least four sugar transferase

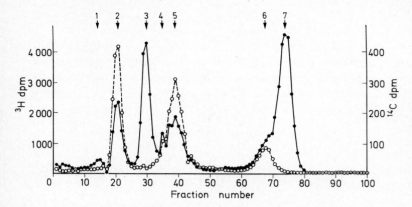

Fig. 4. Electrophoresis of polypeptides of influenza virus WSN grown in MK cells, double labeled with fucose-^{3}H and ^{14}C-amino acid mixture. ●—●, ^{14}C dpm; o—o, ^{3}H dpm. (From COMPANS et al., 1970)

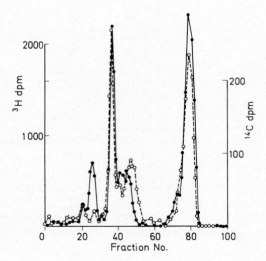

Fig. 5. Coelectrophoresis of polypeptides of influenza virus WSN grown in MDBK cells, labeled with ^{3}H-amino acid mixture, and virions grown in BHK 21-F cells labeled with ^{14}C amino acid mixture. ●—●, ^{3}H dpm; o—o, ^{14}C dpm. (From COMPANS et al., 1970)

enzymes would be required for the synthesis of SV5 glycoproteins. It seems doubtful that the relatively small viral genome would specify so many transferases; it therefore appears that at least part of the carbohydrate structure is specified by host transferases.

That this might be indeed the case is indicated by studies on the WSN line of influenza virus A (COMPANS et al., 1970; SCHULZE, 1970). Fig. 4 shows that this virus contains seven different polypeptides, four of which are glycoproteins, namely the polypeptides 2, 4, 5, and 6. On polyacrylamide gels these glycoproteins have a slightly higher mobility when the virus has been grown in BHK cells than when it has been grown in MDBK cells (Fig. 5). These results suggest that the carbohydrate moiety of these glycoproteins is at least partly specified by the host. These findings are in agreement with previous reports by HAUKENES et al. (1965), LAVER et al. (1966), and LEE et al. (1969), who isolated host specific glycopeptides from influenza virus. Similarly, ROTT et al. (1966) found by means of phytagglutinins host-specific carbohydrates in lipid-containing viruses.

Lipids of SV5 Virions and Host Cell Plasma Membranes

The lipid composition of influenza virus is similar to that of the host cell (FROMMHAGEN et al., 1958; ARMBRUSTER und BEISS, 1958). Host cell lipids labeled radioactively before infection are incorporated into virus particles (WECKER, 1957). Furthermore, KATES and coworkers (1962) found in influenza virus grown in different cell lines host-specific modifications of the lipid composition. Several groups (SPRINGER and TRITEL, 1962; ISACSON and KOCH, 1965; ROTT et al., 1966) were able to detect in influenza virus glycolipid antigens which were also host-specific. On the basis of these findings it is generally accepted that the lipids of the virus particle are derived from the host cell.

We undertook a comparison of the lipid composition of SV5 virions and of the plasma membranes of cells in which they were grown [KLENK and CHOPPIN, 1969 (2)]. This was done for several reasons: (1) to analyze further the structure and composition of SV5 virions and of the plasma membranes of cultured cells; (2) to explore the contribution of the plasma membrane to the virus particle. For these investigations we used primary monkey kidney cells (MK), and continuous lines of bovine kidney cells (MDBK) and of hamster kidney cells (BHK 21-F and HaK).

A number of procedures for isolation of plasma membranes were investigated, and the method selected was a modification of that of WARREN and coworkers (1966) employing fluorescein mercuric acetate. This method gave a good yield of relatively pure plasma membranes with each of the four cells examined [KLENK and CHOPPIN, 1969 (2), 1970 (1)].

With each of the four cell types, a comparison of the lipids of the whole cell with those of the isolated plasma membranes revealed that the plasma membranes have a higher sphingomyelin and cholesterol content than the respective whole cell. These substances are thus major constituents of plasma membranes of cultured cells, as they are in myelin, erythrocyte ghosts, and liver cell plasma membranes. In SV5 virions the sphingomyelin and cholesterol content is as high as it is in the plasma membrane of the host cells.

Table 1 shows the distinctive phospholipid patterns of MK, BHK and HaK plasma membranes and of SV5 grown in these cells. The plasma membrane of the MK cells clearly differs from that of the two hamster cells, which are very similar to each other.

The MK membrane has a lower phosphatidylcholine and a higher phosphatidylethanolamine content than the hamster cells. The phospholipids of the virions grown in different cells also differ, and they show clearly the same distinct phospholipid patterns as the host cell membrane. Thus, in three different cell types with quantitatively distinct phospholipid patterns the SV5 virions closely reflect the lipid composition of the host cell membrane.

The fatty acids of the phospholipids have been examined, too. Significant differences between plasma membranes and virus could not be detected. The fatty acid composition of the medium had a pronounced influence on that of the cells and of the virus.

The data thus far presented indicate that the lipids of SV5 virions closely resemble those of the plasma membrane of the host cell with respect to the distribution of major lipid classes, the individual phospholipids, and fatty acids. The distinctive differences in phospholipid patterns of different membranes are clearly reflected in the virus. There was, however, one exception among the four cell types. In virus grown in MDBK cells, there is a higher concentration of phosphatidylcholine than in the membrane, and these differences appear to be significant. What is interesting about this exception to the general rule is that these values in the MDBK-grown virus are almost identical to those found in the MK cell membrane and MK-grown virus (Table 1). The MK cell produces a very high yield of SV5. This suggested that a relatively high phosphatidylethanolamine and a lower phosphatidylcholine content might be optimal for virus production. Although the virus must, in general, accept

Table 1. *Phospholipid patterns of plasma membranes of monkey kidney, BHK 21-F, and HaK cells and SV5 virions grown in these cells*

Source	Per cent of total phospholipid				
	Sphingomyelin	Phosphatidylcholine	Phosphatidylinositol	Phosphatidylserine	Phosphatidylethanolamine
MK membranes	11.8	32.1	—	17.2	38.8
SV5 from MK cells	12.2	25.2	2.9	17.9	40.3
BHK 21-F membranes	24.2	49.5	10.0	5.1	11.2
SV5 from BHK 21-F cells	30.0	38.5	10.5	5.2	15.6
HaK membranes	24.4	46.8	11.7	5.0	13.0
SV5 from HaK cells	25.8	43.8	8.5	5.0	17.1

Modified from KLENK and CHOPPIN [1969 (2), 1970 (1)].

the lipids of the plasma membrane, in certain instances the viral proteins might exert some limited quantitative selectivity in utilizing the available phospholipids. A similar mechanism for the assembly of viral membranes has been postulated by BLOUGH and his coworkers. [BLOUGH and LAWSON, 1968; TIFFANY and BLOUGH, 1969 (1, 2)].

Glycolipids of the Viral Envelope

The investigations of plasma membrane and viral lipids have been extended to the glycolipids. Glycolipids are composed of sphingosine, fatty acids, and carbohydrate. They are present in a wide variety in the membranes of animal cells and their carbohydrate moiety may serve as antigenic determinants of the cell surface, such as blood group or Forssman antigens (KOSCIELAK et al., 1968; MAKITA et al., 1966; MÅRTENSSON, 1969).

The glycolipids can be divided into two groups: gangliosides, which contain neuraminic acid, and neutral glycolipids, which do

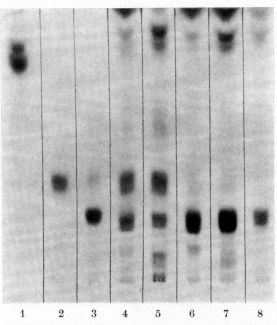

Fig. 6. Thin-layer chromatogram of neutral glycolipids of MK and MDBK cells, plasma membranes, and SV5 virions with reference substances. (1) galactosyl-ceramide, (2) galactosyl-galactosyl-glucosylceramide, (4) MK cells, (5) SV5 grown in MK cells, (6) MDBK cells, (7) MDBK cell plasma membranes (8) SV5 grown in MDBK cells. [From KLENK and CHOPPIN, 1970 (2)]

not contain neuraminic acid. One of the striking findings of our studies is that those cells which contain large amounts of neutral glycolipids contain few gangliosides, and vice versa.

Fig. 6 shows a chromatogram of the neutral glycolipids obtained from MK and MDBK cells and SV5 virions. MDBK cells contain

glucosyl ceramide and N-acetylgalactosaminyl-galactosyl-glucosylceramide; MK cells contain these, and in addition, galactosylgalactosyl-glucosylceramide. Plasma membranes contain the same glycolipids as whole cells, but in 4 to 8 times higher concentrations (Fig. 6, Table 2). The neutral glycolipid content of virus grown in MDBK or MK cells clearly reflects that of the plasma membranes (Fig. 6).
A variety of gangliosides has been identified in the investigated cell lines. There are distinctive qualitative and quantitative differences between the ganglioside of the various cells. BHK 21-F cells

Table 2. *Neutral glycolipid content of MDBK cells, plasma membranes, and SV5 grown in these cells*

Glycolipid	Whole cells	Plasma membranes	SV5 virions
	(µg per 100 mg protein)		
Glu-Cer	160	1300	1550
GalNAc-Gal-Gal-Glu-Cer	510	2200	3000

From KLENK and CHOPPIN [1970 (2)].

contain the most ganglioside in total, predominantly N-acetylneuraminosyl-lactose ganglioside. The other cells have a lower content, but contain in addition two other gangliosides with a more complex carbohydrate chain.
As shown in Table 3, gangliosides were not found in SV5 virions. This finding of the failure of gangliosides of the plasma membranes to be incorporated into the virions is in marked contrast to the results with the neutral glycolipids, cholesterol, and phospholipids of the membrane, which are all incorporated quantitatively into the virion. The neuraminic acid-containing membrane lipids are the only ones absent in the virus. As has been shown earlier, the virus contains no protein-bound neuraminic acid either. One explanation for the absence of neuraminic acid-containing substances in the virus is that the viral neuraminidase is synthesized in infected cells and incorporated into the regions of the cell membrane which become the viral envelope.

Further evidence that the absence of neuraminic acid in SV5 virions is specifically due to viral neuraminidase is that vesicular stomatitis virus (VSV), a membrane-enclosed virus which does not possess neuraminidase activity, does contain the same ganglioside which is found in the host cell membrane (KLENK and CHOPPIN, 1971).

Additional evidence that viral neuraminidase might be responsible for the lack of neuraminic acid in the virion is the fact that neuraminic acid is absent only from the areas of the cell membrane

Table 3. *Ganglioside content of BHK 21-F, HaK, MDBK, and MK cells, plasma membranes, and SV5 grown in these cells*

Cells	Whole cells	Plasma membranes	SV5 virions
	(Glycolipid-bound neuraminic acid, µg per 100 mg protein)		
BHK 21-F	83	320	none detected
HaK	20	—[a]	—
MDBK	9	30	none detected
MK	8	—	none detected

[a] not determined.
From KLENK and CHOPPIN [1970 (2)].

where budding occurs. The rest of the membrane contains neuraminic acid. This has been found chemically and also electron microscopically by staining infected cells with colloidal iron which stains the neuraminic acid residues [KLENK et al., 1970 (2)]. The iron is

Fig. 7. a and b SV5-infected MK cells stained with colloidal iron hydroxide (CIH). The surface of the plasma membrane of the cell, including microvilli (M), is covered with electron-dense iron granules. Budding spherical (arrow) and filamentous SV5 particles are not stained. In b) the iron granules are absent only on the budding virus filament, whereas the adjacent cell membrane is stained up to the base of the filament. Magnification: × 100000.
[From KLENK et al., 1970 (2)]

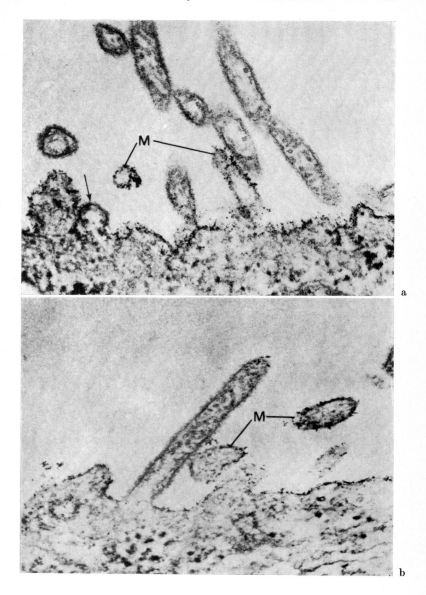

Fig. 7

found everywhere over the surface of the cell except on the budding virus particles (Fig. 7). Thus, one of the specific events that occurs at those areas of membrane which are transformed into viral envelope is the loss of neuraminic acid. It is possible that the localized loss of these residues, which represent a major portion of the negative charge on the cell surface, could play a role in virus assembly, such as facilitation of the incorporation of other viral proteins into the membrane or of the assembly of spikes, or the stimulation of virus budding.

In addition to the significance of glycolipids as structural components of the viral envelope, these substances may be antigenic, e.g. they may possess blood group or Forssman activity (KOSCIELAK et al., 1968; MAKITA et al., 1966; MÅRTENSSON, 1969). Such activities have been detected by immunological means in preparations of myxoviruses and other lipid-containing viruses (ISACSON and KOCH, 1965; ROTT et al., 1966; SPRINGER and TRITEL, 1964). The present demonstration of glycolipids in enveloped viruses, and the finding of glycopeptide host antigens in influenza virus (HAUKENES et al., 1965; LAVER and WEBSTER, 1966; LEE et al., 1969), when coupled with the recent evidence that myxo- and paramyxoviruses do not contain host cell proteins, suggests that these antigens will be carbohydrate rather than protein.

Assembly of Myxovirus Envelopes

The data of the chemical analysis combined with the morphological findings suggest the following — partially still speculative — sequence of events during virus assembly and biosynthesis of the viral envelope. The proteins of the viral envelope are first incorporated into certain areas of the cell membrane. This conclusion is based on several points (CHOPPIN et al., 1971): (1) the very regular arrangement of nucleocapsid beneath only certain areas of the cell membrane implies that it is recognizing some very specific sites rather than associating with the membrane at random; (2) ferritin-labeled anti-viral antibody attaches to viral antigens at sites on the cell membrane which have no nucleocapsid beneath them; and (3) virus-specific hemadsorption occurs on regions of the membrane beneath which no nucleocapsid is seen. Thus, viral hemagglutinin and perhaps other viral proteins appear to be incorporated into the membrane. Then the nucleocapsids align under the regions of the membrane which contain the virus-specific proteins. The nucleocapsids are then enveloped by the spike-covered membrane and released from the host cell as mature virions in a budding process. Despite the morphological similarity of the viral envelope and the cell membrane, the proteins of the plasma membrane have been completely replaced in the viral envelope by virus-specific proteins, whereas the lipids and the carbohydrates show striking host specificities.

Structure of the Viral Envelope

Insight into the organization of the subunits within the virion is furnished by experiments in which fowl plague virus, an influenza virus, has been stepwise degraded. On polyacrylamide gels, the virus yields four major bands which represent polypeptides with molecular weights of 24000, 28000, 49000, and 56000 (Fig. 8). The polypeptides with molecular weights 24000 and 56000 are carbohydrate-free, the others are glycoproteins. Isolated spikes, which can be split from intact virions by treatment with ether and adsorption to red blood cells (HOYLE, 1952; SCHÄFER and ZILLIG, 1955), consist of these glycoproteins (Fig. 8). This means that the hemagglutinin and the neuraminidase, which are separate structures, are formed by these glycoproteins. With all viruses so far investigated the surface projections consist of glycoproteins (COMPANS et al., 1970; SCHULZE, 1970; COMPANS, 1971; CHEN et al., 1971). This is another additional structural feature common to both envelopes and plasma membranes which also have carbohydrate-rich glycoproteins on their outer surface (UHLENBRUCK, 1971).

In order to get insight into the inner layers of the viral envelope, the spikes have been removed from the virion by treatment with bromelin, a proteolytic enzyme from pineapple, and a viral subparticle having a smooth surface without spikes has been isolated (COMPANS et al., 1970). These particles no longer possess hemagglutinin and neuraminidase and they have lost their infectivity. The particles still contain lipids, and in particular the blood group-specific glycolipids. Thus, a lipid-containing layer of the viral envelope, in which the spikes are embedded, has not been degraded by the enzyme treatment.

On the basis of these experiments, one would expect that these spike-free particles no longer contain glycoproteins. This is indeed

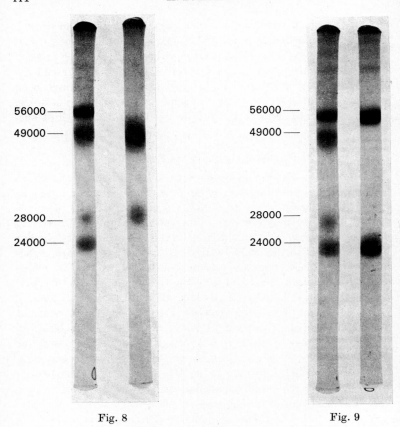

Fig. 8. Polyacrylamide gel electrophoresis of the polypeptides of fowl plague virus. a) complete virions, b) isolated spikes. The molecular weights of the proteins are from SKEHEL and BURKE (1969)

Fig. 9. Polyacrylamide gel electrophoresis of fowl plague virus polypeptides. a) complete virions, b) virions devoid of spikes

the case, as shown in Fig. 9. The polypeptide pattern of these particles contains only the carbohydrate-free polypeptides, the glycoproteins having been lost after the enzyme treatment.

Where are these carbohydrate-free polypeptides localized in the virion ? The one with the higher molecular weight forms together with the viral RNA the internal component, the nucleocapsid, as has been shown for a series of myxoviruses (HASLAM et al., 1970; SCHULZE, 1970; JOSS et al., 1969; ETCHISON et al., 1971; EVANS and KINGSBURY, 1969; BIKEL and DUESBERG, 1969; MOUNTCASTLE et al., 1970). In addition to the nucleocapsid protein, all these viruses have been found to contain a second carbohydrate-free major polypeptide having a molecular weight of 20000 to 30000 with the orthomyxoviruses and about 40000 with the paramyxoviruses. Its exact function is not yet known. However, there are indications that it might be another protein of the viral envelope. Since, in contrast to the spike proteins, it is protease-resistant in the virion, one has to conclude that it is protected from the enzyme action by the virus lipids.

Indeed, in a further degradation step of treating the spikeless particle with detergents, SCHULZE was able to isolate another particle which had lost all lipids but no protein and which was still covered by a thin membrane. Similar results have been reported for vesicular stomatitis virus (CARTWRIGHT et al., 1970). All this indicates that the viral envelope is formed by a central lipid layer which has on its outer surface the spikes consisting of glycoproteins, whereas its inner side is coated by a carbohydrate-free protein. It should be mentioned here that the plasma membrane of the host cells of these viruses contains many different polypeptides. Thus, despite the similarity in morphology and general chemical composition, the viral envelope differs from the plasma membrane in the low number of its structural proteins.

Summary

Myxoviruses are assembled and released by budding from the surface of the host cell. During this budding process the internal component of these viruses is enveloped by a membrane which is continuous with, and morphologically similar to, the plasma membrane of the host cell.

Myxo- and paramyxovirions contain two major carbohydrate-free polypeptides and several glycoproteins. Host cell proteins could not be detected in the virus particles. This indicates that, despite the morphological similarity of the viral envelope and the cell membrane, host proteins have been replaced in the viral envelope by virus-specific proteins.

Unlike the proteins, the majority of the plasma membrane lipids are quantitatively incorporated into the viral envelope. Through glycolipids, host antigens are incorporated into the virion. There are also indications for a host specificity of the carbohydrate moiety of the viral glycoproteins. Since host proteins have not been detected in these viruses, the findings indicate that host antigens in virus particles are not protein but carbohydrate.

Myxoviruses contain neither protein nor lipid-bound neuraminic acid, presumably due to the localized action of the viral neuraminidase on those areas of the cell membrane which become the viral envelope.

Partial degradation of virions by detergents and hydrolytic enzymes yields a variety of subviral particles. The analysis of these particles sheds some light on the virus structure: the ribonucleoprotein of the nucleocapsid is surrounded by the viral envelope. In the envelope two layers can be distinguished: the outer one is formed by the hemagglutinin and the neuraminidase which consist of glycoproteins and appear as radial projections on the surface of the virion. These spikes have their roots in the inner layer of the envelope which appears morphologically as a unit membrane. This membrane is formed by lipid and presumably a single polypeptide species. Because of their relatively simple and clear structure, the envelopes of these viruses promise to be valuable models for the investigation of membrane structure and biogenesis in general.

Acknowledgment

I am indebted to Dr. P. W. CHOPPIN, in whose group and under whose guidance most of this work was carried out. I would also like to thank Drs. R. W. COMPANS and L. A. CALIGUIRI of the Rockefeller University, New York, and R. ROTT and H. BECHT of the University of Giessen for their collaboration.

References

ACHESON, N. H., TAMM, I.: Virology **32**, 128 (1967).
ADA, G. L., GOTTSCHALK, A.: Biochem. J. **62**, 686 (1956).
ARMBRUSTER, O., BEISS, U.: Z. Naturforsch. **13** b, 75 (1958).
BEARD, J. W., BONAR, R. A., HEINE, U., DE THÉ, G., BEARD, D.: Viruses, nucleic acids, and cancer. p. 340—373. Baltimore: The Williams and Wilkins Company 1963.
BEN-PORAT, T., KAPLAN, A. S.: Virology **41**, 265 (1970).
BIKEL, J., DUESBERG, P. H.: J. Virol. **4**, 388 (1969).

Blough, H. A., Lawson, D. E. M.: Virology **36**, 286 (1968).
Bolgnesi, D. P., Bauer, H.: Virology **42**, 1097 (1970).
Burge, B. W., Huang, A. S.: J. Virol. **6**, 176 (1970).
Caliguiri, L. A., Klenk, H.-D., Choppin, P. W.: Virology **39**, 460 (1969).
Cartwright, B., Talbot, P., Brown, F.: J. gen. Virol. **7**, 155 (1970).
Chen, C., Compans, R. W., Choppin, P. W.: J. gen. Virol. **11**, 53 (1971).
Choppin, P. W., Klenk, H.-D., Compans, R. W., Caliguiri, L. A.: From molecules to man, perspectives in virology VII, p. 127. (Pollard, M., Ed.) New York: Academic Press 1971.
— Stockenius, W.: Virology **23**, 195 (1964).
Compans, R. W.: Nature (Lond.) **229**, 114 (1971).
— Choppin, P. W.: Virology **35**, 289 (1968).
— Dimmock, N. J.: Virology **39**, 499 (1969).
— Holmes, K. V., Dales, S., Choppin, P. W.: Virology **30**, 411 (1966).
— Klenk, H.-D., Caliguiri, L. A., Choppin, P. W.: Virology **42**, 880 (1970).
Dales, S., Mosbach, E. H.: Virology **35**, 564 (1968).
Etchison, J., Doyle, M., Penhoet, E., Holland, J.: J. Virol. **7**, 155 (1971).
Evans, M. J., Kingsbury, D. W.: Virology **37**, 597 (1969).
Frommhagen, L. H., Freeman, N. K., Knight, C. A.: Virology **5**, 173 (1958).
— Knight, C. A., Freeman, N. K.: Virology **8**, 176 (1959).
Haslam, E. A., Hampson, A. W., Radiskevicz, E., White, D. O.: Virology **42**, 566 (1970).
Haukenes, G., Harboe, A., Mortensson-Egnund: Acta path. microbiol. scand. **64**, 534 (1965).
Holland, J. J., Kiehn, E. D.: Science **167**, 202 (1970).
Hoyle, L.: J. Hyg. (Lond.) **50**, 229 (1952).
Isacson, P., Koch, A. E.: Virology **27**, 129 (1965).
Joss, A., Gandhi, S. S., Hay, A. J., Burke, D. C.: J. Virol. **4**, 816 (1969).
Kates, M., Allison, A. C., Tyrrell, D. A. J., James, A. T.: Cold Spr. Harb. Symp. quant. Biol. **27**, 293 (1962).
Klenk, H.-D., Choppin, P. W.: (1) Virology **37**, 155 (1969).
— — (2) Virology **38**, 255 (1969).
— — (1) Virology **40**, 939 (1970).
— — (2) Proc. nat. Acad. Sci. (Wash.) **66**, 57 (1970).
— Caliguiri, C. A., Choppin, P. W.: (1) Virology **42**, 473 (1970).
— Compans, R. W., Choppin, P. W.: (2) Virology **42**, 1158 (1970).
— — J. Virol. **7**, 416 (1971).
Koscielak, J., Hakomori, S., Jeanloz, R. W.: Immunochemistry **5**, 441 (1968).
Laver, W. G., Webster, R. G.: Virology **30**, 104 (1966).
Lee, L. T., Howe, C., Meyer, K., Cho', H. U.: J. Immunol. **102**, 1144 (1969).
Makita, A., Suzuki, C., Yorizawa, Z.: J. Biochem. **60**, 502 (1966).
Mårtensson, E.: Progress in the chemistry of fats and other lipids **10**, 367. Pergamon Press 1969.
McSharry, J., Wagner, R. R.: J. Virol. **7**, 412 (1971).
Mountcastle, W. E., Compans, R. W., Caliguiri, L. A., Choppin. P. W.: J. Virol. **6**, 677 (1970).

Mountcastle, W. E., Compans, R. W., Choppin, P. W.: J. Virol. **7**, 47 (1971).
Mussgay, M., Weibel, J.: J. Cell Biol. **16**, 119 (1963).
Olshevsky, U., Becker, Y.: Virology **40**, 948 (1970).
Rott, R., Drzeniek, R., Saber, M. S., Deichert, E.: Arch. Ges. Virusforsch. **19**, 273 (1966).
Schäfer, W., Zillig, W.: Z. Naturforsch. **19** b, 316 (1955).
Schulze, I. T.: Virology **42**, 890 (1970).
Siegert, R., Falke, D.: Arch. ges. Virusforsch. **19**, 230 (1966).
Skehel, J. J., Burke, D. C.: J. Virol. **3**, 420 (1969).
Spear, P. G., Keller, J. M., Roizman, B.: J. Virol. **5**, 123 (1970).
Springer, G. F., Tritel, H.: Science **138**, (1962).
Strauss, J. H., Jr., Burge, B. W., Darnell, J. E.: J. molec. Biol. **47**, 437 (1970).
Tiffany, J. M., Blough, H. A.: (1) Science **163**, 573 (1969).
— — (2) Virology **37**, 492 (1969).
Uhlenbruck, G.: Chimia **25**, 10 (1971).
Warren, L., Glick, M. C., Nass, M. K.: J. Cell Physiol. **68**, 269 (1966).
Wecker, E.: Z. Naturforsch. **12** b, 208 (1957).

Metabolite Carriers in Mitochondrial Membranes: the Ca^{2+} Transport System

ALBERT L. LEHNINGER

*Department of Physiological Chemistry,
The Johns Hopkins University School of Medicine,
725 North Wolfe Street, Baltimore, MD 21205, U.S.A.*

With 1 Figure

In addition to the various electron carriers required in electron transport and the ATP synthetase system participating in oxidative phosphorylation of ADP, the inner mitochondrial membrane contains a number of specific permeases or transport systems which make possible metabolic interplay between the mitochondrial matrix and the surrounding cytoplasm. In this paper the general characteristics of these permease systems will be outlined first. Then our recent work on the properties of the specific carrier involved in mitochondrial Ca^{2+} transport will be described in more detail.

The metabolic traffic across the mitochondrial membrane. Mitochondria are not completely autonomous metabolic units. Actually there is a considerable traffic of metabolites between the cytoplasm and the mitochondrial matrix. All mitochondria, regardless of the cell type, must take in fuels for oxidation, such as pyruvate and fatty acids, and then discharge bicarbonate and CO_2 into the cytoplasm as the end-products of the tricarboxylic acid cycle. Simultaneously, mitochondria must take in phosphate and ADP and discharge ATP as the end-product of oxidative phosphorylation.

In addition to their primary function in respiration and phosphorylation, mitochondria also perform a number of auxiliary metabolic roles, depending on the type of cell. In liver and kidney cells various intermediates of the tricarboxylic acid cycle must enter or leave mitochondria, particularly malate and citrate, which participate in transport of reducing power between the mitochondrial matrix and cytoplasm. Malate also must cross the mitochondrial membrane during gluconeogenesis and anaplerotic reactions. During urea synthesis in liver cells glutamate must enter mitochondria for the delivery of α-amino groups and α-ketoglutarate must leave again after deamination of the glutamate. Moreover, it is well known that the urea cycle enzymes are partitioned between the cytoplasm and the mitochondria, as are various other enzyme systems required for amino acid biosynthesis and degradation and for heme synthesis.

This metabolic traffic across the mitochondrial membrane has two distinctive features. First, only certain metabolites can cross the membrane readily; many others are unable to cross. Secondly, it is kinetically and stoichiometrically integrated so that only the required numbers and ratios of substrate, phosphate, and ADP molecules, for example, flow into and out of the mitochondria in coordination with the metabolic needs of the cytoplasm. The metabolic traffic across the mitochondrial membrane is therefore "computerized" and the molecular basis for the function of the mitochondrial permeases must include appropriate mechanisms for their regulation and integration.

Mitochondrial membrane permeability. The inner mitochondrial membrane, like most natural membranes, appears to be intrinsically impermeable to most polar molecules and particularly to small ions. Thus it is impermeable to sucrose and other sugars and to H^+, OH^-, and Cl^- ions. On the other hand, it is permeable to small neutral molecules such as H_2O, CO_2, and urea, to undissociated monocarboxylic acids such as acetic, butyric, and β-hydroxybutyric acids, and to certain unprotonated bases such as NH_3 and ethylamine. NH_4^+ and CH_3COO^- ions are only indirectly permeant, because at pH 7.0 significant quantities of free NH_3 and CH_3COOH exist and may pass through the membrane by means of unmediated physical diffusion. However, highly charged metabolites, such as the anions of dicarboxylic and tricarboxylic acids, as well as the ionic species of amino acids at pH 7.0, are not readily permeable through most membranes and can penetrate the mitochondrial membrane only through the mediation of specific transport systems.

The penetration of the mitochondrial membranes by specific metabolites can be studied by methods that are quite similar to those used for study of the influx and efflux of permeants in other membrane-surrounded structures, such as erythrocytes or bacterial cells. Thus the rates of influx and efflux of labeled metabolites may be followed, as well as the rates of swelling of mitochondria when placed in isoosmotic solutions of the solute whose penetration is to be measured.

Criteria of mitochondrial permeases. The experimental criteria for verifying the presence of specific mediated transport of a metabolite across the mitochondrial membrane are very similar to those used in study of transport systems in intact cells, such as erythrocytes or bacteria. The transport system must show (1) specificity for the substance transported (2) saturation kinetics (3) susceptibility to specific inhibition (4) specific binding sites for its ligand and (5) it must be genetically determined. Moreover, it may also show a dependence on metabolic energy and a requirement for a specific partner with which it may participate in antiport or symport. Many or all of these criteria have been met in the case of a number of mitochondrial transport systems.

The known mitochondrial permeases. Table 1 lists the known mitochondrial transport systems for which reasonably good evidence exists, together with information on ligand specificity, type of transport promoted, and their specific inhibitors [1]. Perhaps the most thoroughly studied is the ADP-ATP carrier, which normally catalyzes an obligatory stoichiometric exchange of ADP for ATP during oxidative phosphorylation. Its properties have recently been summarized [2, 3]. The ADP-ATP carrier has a very high affinity; K_D for ADP is about 0.5 μM. It occurs to the extent of 0.15 nmoles per mg mitochondrial protein, equivalent to the molar amount of cytochrome c in the inner membrane. The maximum rate of ADP-ATP exchange is about 200 nmoles per mg protein per min, which far exceeds the normal rate of oxidative phosphorylation. It is specifically inhibited by the toxic glycoside atractyloside, which apparently competes with ADP for binding and by bongkrekic acid, which greatly increases its affinity for ADP. The ATP-ATP carrier has not heretofore been extracted in soluble form from mitochondria, nor for that matter have any of the carriers except possibly the Ca^{2+} carrier, to be described presently.

In addition to the carriers listed in Table 1, we have recently adduced evidence for specific carriers for two intermediates of the urea cycle, one which makes possible the entrance of ornithine into the mitochondrial matrix and the other the passage of citrulline from the matrix back into the cytoplasm [4]. Ornithine, which is positively charged at pH 7.0, enters on a carrier system which is dependent on respiration and phosphate; citrulline departs by means of a neutral uniport mechanism.

Table 1. *Mitochondrial transport systems*

System	Type of process	Inhibitor
Pi	$H_2PO_4^-$-OH^- antiport	Mersalyl
ADP	ADP^{3-}-ATP^{4-} antiport	Atractyloside; bongkrekic acid
Dicarboxylate	$Malate^{2-}$-$H_2PO_4^-$ antiport $Malate^{2-}$-$succinate^{2-}$ antiport	n-Butylmalonate; phenylsuccinate
Tricarboxylate	$Citrate^{2-}$-$H_2PO_4^-$ antiport $Citrate^{2-}$-$malate^{2-}$ antiport	Tricarballylate; ethylcitrate
α-Ketoglutarate	α-KG uniport	—
Glutamate	Glu uniport	Avenaceolide
Ca^{2+}	Ca^{2+}-H^+ antiport	La^{3+}
Na^+	Na^+-H^+ antiport	—

Table 2. *Species distribution of mitochondrial transport systems*

System	RLM	Kidney	Heart	Blowfly	17-yr locust
Pi	+	+	+	+	+
ADP	+	+	+	+	+
Dicarboxylate	+	+	+	0	0
Tricarboxylate	+	+	0	0	+
α-Ketoglutarate	+	+	0	0	+
Glutamate	+	0	+	0	+
Ca^{2+}	+	+	+	0	+

Particularly compelling evidence for the individuality of the mitochondrial membrane carriers comes from study of their comparative distribution in mitochondria of different types of cells (Table 2). Thus blowfly flight muscle mitochondria, which are specialized for high rates of respiration and phosphorylation and apparently carry out no other major metabolic function, lack carriers for the tricarboxylic acid cycle intermediates [1, 5], whereas rat liver mitochondria, which actively participate in reductive biosynthesis, amino acid oxidation, and the urea cycle, as well as respiration, possess carriers for a large number of metabolites.

Integration of carrier action. The mitochondrial transport systems function in an integrated network. Several factors are involved in such integration. First, it is noted that some of them carry out stoichiometric antiport processes, which bear a relationship to their biological function. Thus, the ADP carrier allows one ATP molecule to leave the matrix for each ADP entering; this is of course the ratio in which these nucleotides participate in oxidative phosphorylation. In addition, the glutamate carrier may function in stoichiometric antiport with the α-ketoglutarate carrier during the delivery of amino groups to the matrix for urea synthesis.

The second factor which makes possible coordination of the various carriers is the fact that electron transport can drive the transport of H^+ from the matrix to the outside. For each pair of electrons traversing each of the three energy-conserving sites of the chain, two H^+ ions may be transported out, by mechanisms still unknown, leaving the matrix alkaline and the outside relatively acid and thus creating a pH gradient. This pH gradient can be converted into a gradient of electrical potential by the specific phosphate carrier, which promotes an exchange of external phosphate^{-1} for internal OH^- ions [1, 6, 7], as is shown in the top portion of Fig. 1, resulting in electrogenic entry of phosphate into the matrix. Since phosphate is a required component in oxidative phosphorylation and since phosphate is also required for and may participate in the coupled transport of tricarboxylic and dicarboxylic acids across the membrane, phosphate transport plays a central role in the integration of flows across the mitochondrial membrane. Fig. 1 shows how the electrogenic entry of phosphate makes possible the entry of one molecule of ADP, bearing three negative charges, and the exit of one molecule of ATP, which bears four negative charges at pH 7.0. Thus the law of electroneutrality ensures the correct stoichiometry of transport of Pi, ADP, and ATP during oxidative phosphorylation [2].

The Ca^{2+} carrier: general properties. For some years it has been known (for reviews, see [8, 9]) that mitochondria of various animal tissues can accumulate Ca^{2+} during respiration, particularly in the presence of phosphate; 2 Ca^{2+} ions are accumulated per pair of electrons passing through each energy-conserving site of the respiratory chain. Simultaneously, phosphate is also accumulated. Very large amounts of Ca^{2+} and phosphate may be accumulated in this manner, leading to deposition of large electron-dense deposits of tricalcium phosphate in the matrix. Ca^{2+} also stimulates respiration of mitochondria stoichiometrically, to even higher rates than yielded by ADP. Indeed, the affinity of rat liver mitochondria is higher for

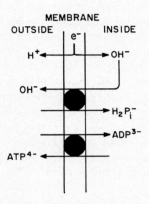

Fig. 1. Entry of ADP^{3-} and exit of ATP^{4-} coupled to the energy-dependent electrogenic entry of phosphate

Ca^{2+} than for ADP, so that when both are added simultaneously, respiration-dependent accumulation of Ca^{2+} takes place in preference to oxidative phosphorylation of ADP [10]. The primacy of Ca^{2+} uptake over oxidative phosphorylation is observed in mitochondria of all vertebrate tissues and suggests that energy-linked Ca^{2+} uptake is a basic and fundamental attribute of mitochondria, comparable in importance to oxidative phosphorylation [9].

The very high affinity of mitochondria for Ca^{2+}, their specificity for Ca^{2+}, Sr^{2+}, and Mn^{2+} uptake, and the experimental dissociation of Ca^{2+} transport from electron transport, strongly suggested the occurrence of a specific Ca^{2+} carrier in the membrane, one catalyzing a Ca^{2+}-H^+ antiport [9]. Moreover, the specific inhibition of Ca^{2+}

transport by La^{3+} and other rare earth cations [cf. 11] and by ruthenium red [12] also is consistent with the occurrence of a Ca^{2+} carrier.

Evidence for a Ca^{2+} carrier was further strengthened by the finding of REYNAFARJE and LEHNINGER [13] that rat liver mitochondria contain two classes of respiration-independent Ca^{2+} binding sites, one of very high affinity but small in number, the other of low affinity but large in number. The high affinity Ca^{2+} sites had very similar specificity for Ca^{2+}, Sr^{2+}, and Mn^{2+}, and were blocked by La^{3+} [11], which strongly indicated that they represent the ligand binding sites of the Ca^{2+} carrier. This evidence was in turn reinforced by the results of a survey of the distribution of Ca^{2+} transport activity and the distribution of high-affinity Ca^{2+} binding capacity in mitochondria isolated from a large variety of cell types. Mitochondria of all vertebrate tissues, including those from birds, reptiles, and amphibia, not only possessed the capacity for Ca^{2+} uptake but also high-affinity Ca^{2+} binding sites [14]. Among lower invertebrates, mitochondria from *Neurospora* and yeasts lack both capacities, but those from amoeba had them in high degree. Blowfly flight muscle mitochondria lack the capacity for Ca^{2+} transport and for high-affinity Ca^{2+} binding [15], but mitochondria from the 17-year cicada possess both [16]. Without exception, all species of mitochondria possessing Ca^{2+} transport capacity also possess high-affinity Ca^{2+} binding sites. Conversely, those lacking Ca^{2+} transport activity lack Ca^{2+} binding sites. We have therefore concluded that the high-affinity Ca^{2+} binding sites of mitochondria are the ligand-binding sites of a genetically-determined Ca^{2+} transport system.

Extraction and purification of a Ca^{2+} binding protein. Using the capacity for high-affinity Ca^{2+} binding as an identifying characteristic of the Ca^{2+} carrier, we have found it possible to extract in soluble form a fraction capable of high-affinity Ca^{2+} binding from rat liver mitochondria exposed to osmotic shock in distilled water [17]. The Ca^{2+} binding activity, which was assayed by microequilibrium dialysis, was found to be heat-labile, stable to freezing and thawing, and to be non-diffusible through cellophane. By means of gel filtration on Sephadex columns the Ca^{2+} binding activity was found to have a Stokes' radius corresponding to a spherical molecule of particle weight 150,000 daltons. The solubilized material bound Ca^{2+}, Sr^{2+}, and Mn^{2+} but not Mg^{2+} and showed an affinity for Ca^{2+}

somewhat lower than that of intact mitochondria. Ca^{2+} binding by the soluble factor was blocked by La^{3+} [17].

More recently, work by Drs. GOMEZ-PUYOU and TUENA in our laboratory has resulted in considerable purification of the Ca^{2+} binding activity by precipitating it with low concentrations of ammonium sulfate, which allowed nearly complete yields and purification of about 100-fold. However, following this salt precipitation procedure, the Ca^{2+} binding factor became completely insoluble in water and was therefore assayed for activity as a suspension following sonication. Extraction of the insoluble factor with acetone yielded an insoluble protein residue, inactive by itself, which was found to give a single sharp band on gel electrophoresis in the presence of SDS and excess mercaptoethanol; its molecular weight was found to be 57,000. The acetone solution on chromatography on methylated Sephadex yielded an acidic lipid fraction which contained nearly all the Ca^{2+} binding activity.

Although it is clear that the Ca^{2+} binding factor is complex, consisting of a protein and one or more acidic acetone-soluble components, it exhibits many of the properties of the Ca^{2+} binding sites of intact mitochondria and may represent at least a portion of the Ca^{2+} transport system. Attempts are under way to reconstitute energy-linked Ca^{2+} transport in transport-negative mitochondrial "ghosts" as well as passive Ca^{2+} transport in synthetic phospholipid bilayers between two aqueous phases. In the latter experiments lowering of membrane resistance by Ca^{2+} and a high transference number for Ca^{2+} will serve as experimental criteria for Ca^{2+} transport activity.

Biological significance of mitochondrial Ca^{2+} transport. Our observations that the mitochondria of all vertebrate tissues (as well as the mitochondria from many invertebrate organisms) possess the capacity for reversible, energy-linked accumulation of Ca^{2+} with very high affinity, as well as the fact that Ca^{2+} uptake takes precedence over oxidative phosphorylation of ADP for respiratory energy, strongly suggest that Ca^{2+} transport by mitochondria plays a very fundamental role in the physiology of many types of cells [9]. The concentration of Ca^{2+} in the cytoplasm is now known to be a crucially important factor in the regulation of a number of important cellular functions, such as the contraction of the actomyosin system of muscle, the Na^+ and K^+ movements involved in the action potential of neurons, the function of microtubules, the regulation of respiratory rate in some muscles, the regulation of the activity of the Na^+, K^+-stimulated ATPase, the regulation of phospholipase activity, the function of the cyclic AMP system (for example, in the stimulation of phosphorylase b kinase), and in the regulation of cytoplasmic communication between epithelial cells. Elsewhere we have summarized evidence that in heart muscle, mitochondria rather than the sarcoplasmic reticulum, plays a major role in segregating free Ca^{2+} during the relaxation phase of the contraction-relaxation cycle [9]. Of especial interest, however, is the possible role of mitochondria in calcification processes, for which we have provided a working hypothesis and theory [9]. It now appears very likely that the mitochondria are the sites at which amorphous tricalcium phosphate is precipitated in the form of micropackets, at the expense of electron transport energy, as a necessary first stage in the formation of hydroxyapatite in normal and pathological calcification processes [9].

Summary

The mitochondrial membrane(s) contain several specific metabolite transport systems, i.e. for phosphate, ADP and ATP, dicarboxylates, tricarboxylates, and amino acids, whose activity is essential for the metabolic and compartmental integration of cell activities. Most of these transport systems may be coupled stoichiometrically to phosphate transport, which is in turn coupled to electron transport. The respiration-dependent transport of Ca^{2+} into mitochondria depends on a specific Ca^{2+} transport system, which is specific for Ca^{2+}, Sr^{2+}, and Mn^{2+}, inhibited by La^{3+}, has a very high affinity for Ca^{2+}, and which is present in mitochondria of all vertebrate tissues, but lacking in mitochondria of yeast, *Neurospora*, blowfly muscle, and some plant tissues. The presence of this system can be detected by the presence of high-affinity binding sites for Ca^{2+}. A factor capable of high-affinity Ca^{2+} binding has been extracted from rat liver mitochondria and purified. It appears to contain a homogeneous insoluble protein and one or more acidic lipids. Its affinity for Ca^{2+}, its specificity, and inhibition by La^{3+} closely resemble those of the Ca^{2+} transport system of the intact membrane. Ca^{2+} transport by mitochondria, which takes primacy over oxidative phosphorylation, is involved in many aspects of cell physiology, particularly in regulation of metabolism, and in calcification.

References

1. CHAPPELL, J. B.: Brit. med. Bull. **24**, 150 (1968).
2. KLINGENBERG, M.: In: Essays in biochemistry, Vol. 6, p. 119. New York: Academic Press 1970.
3. VIGNAIS, P. V., DUEE, E. D., COLOMB, M., REBOUL, A., CHERUY, A., BARZU, O., VIGNAIS, P. M.: Bull. Soc. chim. Biol. **52**, 471 (1970).
4. GAMBLE, J. G., LEHNINGER, A. L.: Fed. Proc. 30, (1971); — J. biol. Chem., 1971 (in press).
5. VAN DEN BERGH, S. G., SLATER, E. C.: Biochem. J. **82**, 362 (1962).
6. TYLER, D. D.: Biochem. J. **111**, 665 (1969).
7. FONYO, A.: Biochem. biophys. Res. Commun. **32**, 624 (1968).
8. LEHNINGER, A. L., CARAFOLI, E., ROSSI, C. S.: Advanc. Enzymol. **29**, 259 (1967).
9. — Biochem. J. **119**, 129 (1970).
10. ROSSI, C. S., LEHNINGER, A. L.: J. biol. Chem. **239**, 3971 (1964).
11. LEHNINGER, A. L., CARAFOLI, E.: Arch. Biochem. **143**, 506 (1971).
12. MOORE, C. L.: Biochem. biophys. Research Commun. **42**, 298 (1971).
13. REYNAFARJE, B., LEHNINGER, A. L.: J. biol. Chem. **244**, 584 (1969).
14. CARAFOLI, E., LEHNINGER, A. L.: Biochem. J. **122**, 681 (1971).
15. — HANSFORD, R., SACKTOR, B., LEHNINGER, A. L.: J. biol. Chem. **246**, 964 (1971).
16. HANSFORD, R.: Biochem. J. **121**, 771 (1971).
17. LEHNINGER, A. L.: Biochem. biophys. Res. Commun. **42**, 312 (1971).

Membrane Phospholipid Metabolism During Cell Activation and Differentiation

E. FERBER

Max-Planck-Institut für Immunbiologie, D-7800 Freiburg-Zähringen

With 7 Figures

1. Introduction

Changes in cell function involved in cell-mediated immunological reactions, contact, differentiation, cytolytic reactions and responses to extracellular regulators, are probably often triggered by alterations of the outer membrane.

Since phospholipids are integral components of cell membranes but turn over rapidly — at least in their fatty acid moieties — it is reasonable to suspect that they participate in these processes. Moreover, recent investigations have shown that hydrophobic interactions between the fatty acid moieties of membrane phospholipids and apolar parts of membrane proteins are most important for the functional integrity of biological membranes [1—8]. Thus, treatment with phospholipase A (formation of lysophosphatides within the membrane) leads to significant changes in membrane optical activity and function, as reported by WALLACH [3]. Further evidence for this comes from the solubility of membranes in organic solvents [1] and the effects of detergents. On the other hand, removal of the hydrophilic parts of erythrocyte membrane phospholipids with phospholipase C produces only minor changes in membrane structure [8]. Moreover, it is also possible that some membrane phospholipids may actually occur as bilayers within defined membrane areas [2]. In such areas transition from monoacylphosphatides (lysophosphatides) to diacylphosphatides (such as lecithin) or vice versa may induce drastic structural changes. Evidence for this has been provided by SAUNDERS [9] in his study of the properties of lysolecithin/lecithin mixtures: while diacylphosphatides formed lamellar aggregates and monoacylphospholipids spheric micelles, mixtures of the two yield half-open micelles, freely accessible to water and prone to form helical structures. This direct effect of lysophosphatides on the structure of artificial lipid membranes is paralleled by their action on biological membranes, e.g. cytolysis [10, 11] at high concentrations and fusion [12, 13] or stimulation [14, 15] of cells at lower levels. Furthermore, lysophosphatides play an important part as intermediates of phospholipid metabolism. I will first summarize this role and then turn to our own studies in 1) differentiating amoebae (transition from free-living cells to aggregation competent), 2) erythrocytes (especially during ageing), and 3) lymphocytes during the early stages of activation.

2. Metabolism of Fatty Acid Moieties of Phospholipids

An important feature of diacylphospholipids, compared with neutral lipids, is their high content of unsaturated fatty acids

Table 1. *Fatty acid composition of triglycerides and lecithin in rat liver plasma membranes. According to* Wood [76] *and* Henning et al. [77]

	Palmitic 16:0	Stearic 18:0	Oleic 18:1	Linoleic 18:2	Linolenic 18:3 +20:1	Arachidonic 20:4
Triglycerides (Wood, 1970)	32.5	6.5	33.0	10.6	1.0	1.0
Lecithin (Henning u. Stoffel, 1970)	28.3	25.3	9.3	23.7	—	12.8

(Table 1) and an asymmetric distribution of saturated and unsaturated fatty acids, with the saturated fatty acids predominantly found in position 1 of the glycerophosphatide molecule and the unsaturated ones in position 2 [16—21]. During the past decade several workers have searched for mechanisms which could explain this distribution, but several possibilities are still under consideration: the asymmetry is introduced 1) during the first step of *de novo* synthesis, i.e. the acylation of glycerol-3-phosphate; 2) distinct diglycerides are selected during the transfer of choline phosphate; 3) additional enzyme reactions are involved. Possibility 2) is unlikely because the CDP-choline transferase has the same affinity for diglycerides with asymmetric fatty acid distribution [22].

2.1 Diacyl-monoacyl-phosphatide Cycle

A more attractive hypothesis is that of LANDS [23—25] postulating a rearrangement of the fatty acid moieties by de- and reacylation processes (Fig. 1). This hypothesis envisages the deacylation catalyzed by a phospholipase A. However, the well-known snake venom phospholipases have a high specificity for the ester bond in

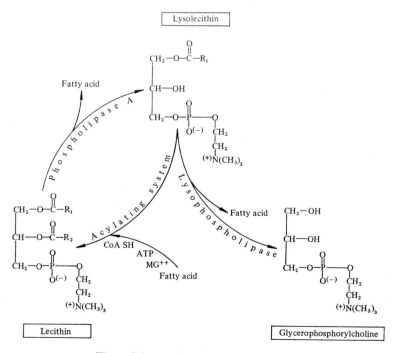

Fig. 1. Scheme of lysolecithin metabolism

position 2 of diacylphosphatides. Known phospholipases exhibit no special affinity for specific fatty acids, but lysolecithin-acyltransferase preferentially transfers higher unsaturated fatty acids to position 2 [24, 26]. According to STOFFEL et al. [27], this specificity of lysolecithinacyltransferase is most marked when the fatty acids are added in

Table 2. *Acyltransferase activities with linoleoyl-CoA and 1-acylglycero-3-phosphorylcholines in pig liver microsomes.* According to BRANDT and LANDS [28]

Lysolecithin	12:0	14:0	16:0	18:0	18:1	18:2
$\dfrac{\text{n moles}}{\text{mg} \times \text{min}}$	18	20	15	14	21	26

Table 3. *Lysophosphatide acyltransferase activities in rat liver microsomes.* According to LANDS et al. [25] [$nmoles \times mg^{-1} \times min^{-1}$]

Acceptor → Acyl-CoA ↓	1 HO-2 3 └P-N 1-Acyl-GPC	┌OH └P-N 2-Acyl-GPC	HO─┐ └P 1-Acyl-GP
14:0	6.3	11.0	6.7
16:0	4.2	11.0	5.8
18:0	2.5	6.7	3.3
18:1	16.0	2.1	24.0
18:2	19.0	3.7	26.0

GPC = glycero-3-phosphorylcholine;
P-N = phosphorylcholine;
P = phosphate.

their acyl-CoA form as saturated and unsaturated pairs. However, although these transferases have a high specificity for unsaturated fatty acids, they do not preferentially acylate saturated lysolecithins [26,28]. Thus, lysolecithins containing different fatty acids from $C_{12:0}$ to $C_{18:2}$ are acylated at similar rates with linoleic acid (Table 2). These experiments also showed an unexpected tendency of unsaturated lysolecithins to incorporate this unsaturated fatty acid [28]. Thus, if this mechanism were the only one involved, it would lead to lecithin species containing *two* unsaturated fatty acids; such species, however, are found only in small amounts. The situation has now been somewhat clarified: thus, VAN DEENEN [29, 30] and LANDS [25] have shown that naturally occurring lysolecithins exist in two isomeric forms, i.e. 1-acyl- and 2-acyl-compounds, which are formed by specific phospholipases, (phospholipase A_1 and phospholipase A_2, respectively [31, 32]) (see

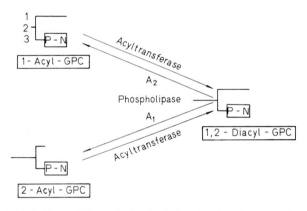

Fig. 2. Metabolism of isomeric lysolecithins. GPC = glycero-3-phosphorylcholine. P-N = phosphorylcholine

Fig. 2) and are specifically acylated by the corresponding lysolecithin-acyltransferases: [1-acyl]-lysolecithins react preferentially with unsaturated and [2-acyl]-lysolecithins with saturated fatty acids. This specificity does not depend on the degree of unsaturation of the lysolecithins and relates only to these two positions on the glycerol moiety of the phosphatides (Table 3).

2.2 Phosphatidic Acid

The simple formulation given above cannot be complete because of the fact that an asymmetric distribution of fatty acids is found already in phosphatidic acid [33, 34]. This could, of course, also be explained by appropriate acylation of the lyso-compound [27, 35,

36]; it is not yet clear, however, whether lysophosphatidic acids exist as intermediates in the *de novo* synthesis of phospholipids. Futhermore, the acylation of α-glycerophosphate yields an asymmetric distribution of fatty acids in phosphatidic acid only when intact cells (or tissue slices) are used and not with microsomal fractions [37, 38].

2.3 Other Mechanisms

Mechanisms other than those mentioned above may be involved. For instance, the acylation of dihydroxyacetonephosphate found by AGRANOFF and collaborators [39—43] could account for some of the otherwise unexplained findings, because a) the enzyme involved appears to have a specificity for saturated fatty acids, thus yielding saturated acyldihydroxyacetonephosphate; b) the reduction of the keto group, in the presence of mitochondria, yields lysophosphatidic acid. However, other possibilities, such as the acylation of monoglycerides to diglycerides [44—46], should also be considered.

3. Cytolytic Action of Lysophosphatides

The formation of lysophosphatides and their conversion to diacylphosphoglycerides is of biological significance. Because lysophosphatides are membranolytic and hence cytotoxic, it is not surprising that their content in cellular membranes is under the control of certain enzymes, namely lysolecithin-acyltransferase and lysophospholipase [47—50]. It is also important to know what are the activities of these membrane-bound enzymes in cells of different function, particularly since these tend to be high compared with other lipid-metabolizing enzymes.

Most of the results referred to have been obtained with liver. However, this may not be the ideal tissue for such investigations for the following reasons: a) it contains many cells other than hepatocytes, e.g. macrophages and endothelial cells; b) the plasma membrane of hepatocytes consists of at least two parts, i.e. bile fronts and blood fronts, which have different functions. In contrast, our own work has involved rather homogeneous cell populations with distinct functions and high proportions of plasma membrane, namely erythrocytes, lymphocytes and certain amoebae.

3.1 Erythrocytes

These cells have two advantages: 1) their plasma membranes can be easily isolated; 2) their lysolecithin-acyltransferase and the lysophospholipase are the only enzymes active in phospholipid metabolism; the cells lack *de novo* phospholipid synthesis and phospholipase-A activity [47—50]. It is generally accepted that during ageing of mammalian erythrocytes glycolytic enzyme activities decline [51—53]. Whether these changes lead to a decreased ATP-level is not certain [54], but even if one assumes that the ATP-level in old erythrocytes declines critically with cell age, this does not explain their elimination from the circulation because, under physiological conditions, old red cells do not die intravascularly but are phagocytosed by reticuloendothelial macrophages. The question now arises as to how spleen macrophages recognize senescent cells. It is unlikely that macrophages can distinguish intraerythrocytic enzyme activities or ATP levels, but they may well be able to sense changes in the outer membrane. For example, the surface charge of old erythrocytes is markedly reduced (DANON [55, 56]) and this might possibly permit phagocytosis of these cells by macrophages, but this is uncertain.

We have investigated the role of membrane-bound lysolecithin-acylating enzymes and lysophospholipase in the ageing process, focusing on their possible protective function against exogenous lysolecithin [57—61]. This question is particularly pertinent since macrophages possess relatively large amounts of phospholipase A [62]. Erythrocytes of differing age can be prepared according to the method of BORUN [63] and RIGAS [64], exploiting the different densities of young and old cells. Fig. 3 shows such a separation in a gradient of albumin; young cells have a low density and are therefore found in the upper layer after centrifugation, while older cells pellet to the bottom of the gradient; subsequent determination of the lysolecithin-acylating activity in the two fractions (Table 4) shows the acylating activities in older cells to be 70% less than in young cells. Our calculation does not take account of the fact that the bottom fraction contains some critically damaged prelytic cells, which would probably make the real specific activities lower than estimated. Not all enzyme activities decrease during ageing, e.g. the Na-K-activated ATPase was not diminished. Further reduction of lysolecithin conversion may be due to the low ATP levels in old

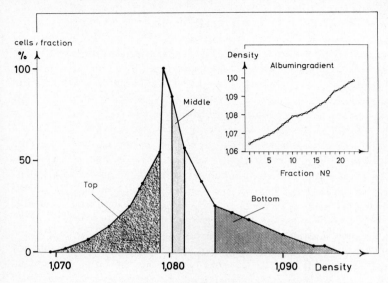

Fig. 3. Distribution of human erythrocytes in an albumin gradient

Table 4. *Activity of the acylating enzyme system in erythrocytes of different density*

$$\frac{\text{Mol} \times 10^{-20}}{\text{cell} \times \text{min}}$$

		Mean	%
Top	258	235	100
	240		
	215		
	225		
Bottom	73	78	33
	83		
	78		
	77		

erythrocytes. In model experiments with reconstituted ghosts we found that the physiological intracellular ATP level (1 μM/ml) is critical for the activation of long-chain fatty acid through the acyl-

CoA-ligase reaction [60, 61, 65, 66]; the intracellular ATP content could thereby influence the structure and function of the outer membrane. In summary, our data suggest that a decrease in lysolecithin-acylating activity accounts for the elimination of old erythrocytes because exogenous lysolecithin, produced by macrophages, cannot be effectively eliminated. This mechanism would, of course, be favored if close contact between the two cell types is fostered by the reduced surface charge of old erythrocytes.

3.2 Amoebae

Our experiments with amoebae of *Dictyostelium discoideum* were undertaken mainly to study biochemical events during morphogenesis: these cells transform, 9 h after consumption of food bacteria, from the free, vegetative form to an aggregation-competent state whence develop tissue-like cell colonies. During this development, surface antigens (mainly polysaccharides), which are not present in vegetative amoebae, appear, while others disappear (GERISCH [67—69]). We had found earlier that homogenates of these amoebae exhibit high activities of membrane-bound phospholipases [70] so that it was of interest to learn more about their biological significance. Accordingly, we have investigated the *de novo* synthesis of phospholipid of intact vegetative and aggregation-competent cells by measuring the incorporation of choline into the lecithin fraction; however, we found no differences between these two stages of development. We therefore followed the incorporation of ^{14}C-labeled long chain fatty acids with and without lysolecithin as the acceptor. As shown in Fig. 4, the incorporation of oleic acid increases significantly as a function of lysolecithin concentration only in the vegetative form. At optimal substrate concentration the incorporation was 3 to 4 times higher in vegetative cells than in aggregation-competent ones. That the increased incorporation of oleic acid is not caused by enhanced permeation due to the detergent properties of lysolecithin is shown by the following results: I) lipid synthesis is restricted to lecithin; II) the kinetics of incorporation are similar whether one uses labeled fatty acids or labeled lysolecithin; III) experiments with double-labeled lysolecithin (label in both phosphate and fatty acid moieties) show that the molecule is metabolized without rearrangement; IV) the distribution of incorporated fatty acid in the lecithin molecule implicates the added lysolecithin as an acceptor, as shown in Table 5, because exogenous lysolecithin causes a shift of the fatty acid distribution to position 2 of the molecule, which is consistent with acylation of [1-acyl]-lysolecithin.

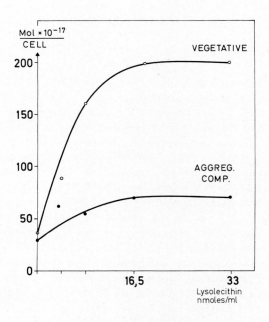

Fig. 4. Incorporation of [1-^{14}C] oleic acid into lecithin of *Dictyostelium discoideum*. 7×10^6 cells were incubated for 10 min, 23 °C in 3 ml 0.016 M phosphate buffer, pH 6.0, containing 35 nmoles [1-^{14}C] oleic acid and 0 to 100 nmoles lysolecithin

The increase in the incorporation rates in vegetative cells cannot be due to active phagocytosis because the cells were washed and incubated without bacteria, and phagocytosis either of bacteria or inorganic particles does not significantly change the acylating activity. Also, the aggregation-competent cells have not lost their phagocytosing capability prior to forming colonies.

More direct evidence that the biochemical changes described are related to the transition from free-living cells to cells which acquire the capability to establish cell contacts comes from experiments with the aggregation-incompetent mutant M_2 aggr. 50—4. As

Table 5. *Positional distribution of ^{14}C-oleic acid incorporated into lecithin of Dictyostelium discoideum. Incorporation of [1-^{14}C]-oleic acid as in Fig. 4. Separated lecithin was degraded using snake venom phospholipase A*

Added Lysolecithin	Vegetative cells		Aggregation-competent cells	
[n moles/ml]	1-Position %	2-Position %	1-Position %	2-Position %
0	29.5	70.5	26.4	73.6
17	9.3	90.7	10.1	89.9
33	3.9	96.1	9.2	90.8

Table 6. *Incorporation of [1-^{14}C]-oleic acid into lecithin of Dictyostelium discoideum wild type and M_2aggr 50-4 mutant. 7×10^6 cells were incubated for 10 min, 23°C in 3 ml 0.016 M phosphate buffer, pH 6.0, containing 68 nmoles ^{14}C-oleic acid and 0—40 nmoles lysolecithin*

Added Lysolecithin	$\left[\frac{10^{-17} \text{ Mol}}{\text{Cell} \times \text{min}}\right]$			
	Wild type		Mutant	
[n moles/ml]	Veg. $\xrightarrow{9\,h}$	Aggr. Comp.	Veg. $\xrightarrow{9\,h}$	Aggr. Incomp.
0	3.5	3.5	10.2	17.2
13	24.0	9.5	39.4	42.5

depicted in Table 6, the acylating activity of this mutant is similar to that of the vegetative cells of the wild type, or even higher, also 9 h after consumption of bacteria, i.e. during the period when the wild type becomes aggregation-competent.

As mentioned above, amoebae of *Dictyostelium discoideum* contain large amounts of phospholipase A and we have shown that this enzyme is also active in intact cells. Fig. 5 shows that degradation of endogenous lecithin is induced by inhibiting the acylating processes with para-chloromercuribenzoate, acylation of lysophosphatides being presumably more rapid than the reverse processes. The degrading velocities are different in the two cell forms, the vegetative cells showing a 2.5 times higher breakdown rate.

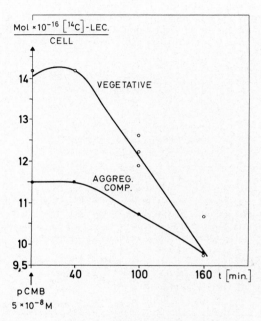

Fig. 5. Degradation of endogenous ^{14}C-lecithin in *Dictyostelium discoideum*. 1.4×10^7 cells were incubated for 20 min at 23 °C in 3 ml 0.016 M phosphate buffer, pH 6.0, containing 35 nmoles [1-^{14}C]-oleic acid and 50 nmoles lysolecithin. Then pCMB (p-chloro-mercuribenzoate) was added to a final concentration of 5×10^{-8} M

3.3 Lymphocytes

Since lymphocytes differentiate from small, resting cells to highly active blasts after binding of stimulating agents on the cell surface, as specific antigens or unspecific stimulants like phytohemagglutinin (PHA), they seemed to us to be very suitable for the study of changes in plasma membrane metabolism immediately after stimulation. Indeed, functional alterations of the plasma membrane, e.g. changes in the permeability of uridine (HAUSEN [71, 72]), are known to occur soon after lymphocyte activation, and our own work points to early changes in phospholipid metabolism. We first compared the incorporation of choline and long-chain fatty acids into lecithin in PHA-stimulated lymphocytes, and found that the fatty acid incorporation started earlier and exceeded that of choline about 20-fold [15]. Equally striking was the distribution

Table 7. *Positional distribution of radioactive fatty acids incorporated into lecithin of rabbit lymphocytes. 1. Incorporation into intact cells: 2×10^7 lymphocytes were incubated for 2 h at 37 °C in 1 ml Eagle's medium containing 9 nmoles [1-^{14}C]-oleic acid or [1-^{14}C]-linoleic acid and 5 nmoles lysolecithin as indicated. 2. Incorporation into microsomes: 100 to 200 µg microsomal protein were incubated for 20 min at 37 °C in 1 ml 0.1 M phosphate buffer, pH 7.5, containing 65 nmoles CoA, 10 µmoles ATP, 10 µmoles $MgCl_2$, 10 nmoles [1-^{14}C]-oleic acid or [1-^{14}C]-linoleic acid and 50 nmoles lysolecithin, as indicated. 3. Separated labeled lecithin was degraded using snake venom phospholipase A*

Fatty acid	Added Lyso-lecithin	Intact Lymphocytes		Microsomes	
		1-Position %	2-Position %	1-Position %	2-Position %
Oleic acid	+	76	24	5.2	94.8
	−	78	22	19.4	80.6
Linoleic acid	+	52	48	8.2	91.8
	−	54	46	21.7	78.3

of the incorporated fatty acids in the lecithin molecule (Table 7). In contrast to the usual pattern, e.g. as in amoebae, unsaturated fatty acids such as oleic and linoleic acid localize predominantly in position 1. As already noted, this could arise from acylation of endogenous [2-acyl]-lysolecithin, but this distribution occurred only with intact cells, not microsomes, and we therefore presume that pathways other than the acylation of lysolecithins are involved. In a second series of experiments we analysed some phospholipid-metabolizing enzymes in subcellular particles of stimulated lymphocytes and found particularly interesting changes in the lysolecithin-acyltransferase. In control samples we find lysolecithin-acyltransferase in the crude, large-granule fraction (mitochondria, lysosomes) and in the microsomal fraction (Fig. 6). However, very shortly after stimulation with PHA the lysolecithin-acyltransferase activity rises, but exclusively in the microsomes. The further fractionation of microsomal membranes into plasma membrane and endoplasmic reticulum was carried out by the method of WALLACH [73, 74]. This

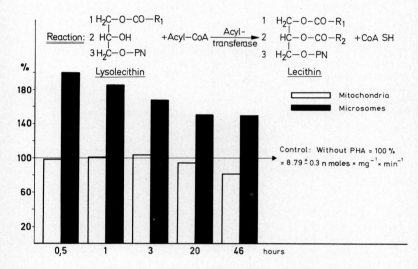

Fig. 6. Stimulation of rabbit lymphocytes with phytohemagglutinin (PHA). Specific activity of lysolecithin acyltransferase in mitochondria and microsomes. (Cultivation 0.5—46 h.) Conditions of the test system: 1 ml 0.1 M phosphate buffer, pH 7.5, contained 50 nmoles 1-acyl [^{14}C]-glycero-3-phosphorylcholine, 50 nmoles oleoyl-CoA and 30—80 µg protein of microsomes or mitochondria; incubation 10 min, 37 °C

method utilizes cell rupture by cavitation of an inert gas and subfractionation of the microsomal vesicles according to their different internal fixed charges, the Donnan ions being titrated with Mg^{++} and the vesicles separated in an osmotically inactive gradient of dextran. With lymphocytes, we obtain two microsomal membrane fractions, 75% of the total consisting of plasma membrane fragments, perhaps contaminated with Golgi membranes, the remainder of endoplasmic reticulum. The characterization and identification of these fractions is based on marker enzymes, such as 5′-nucleotidase for plasma membranes.

As shown in Fig. 7, lysolecithin-acyltransferase localizes almost exclusively in the plasma membrane fraction and increases mark-

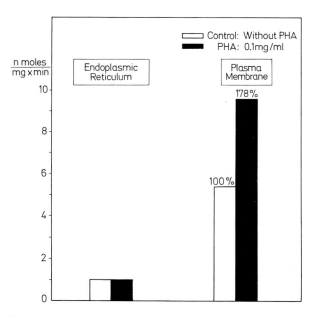

Fig. 7. Stimulation of rabbit lymphocytes with phytohemagglutinin (PHA). Specific activity of lysolecithin acyltransferase in endoplasmic reticulum and in plasma membranes (cultivation 1 h). Conditions of the test system: see Fig. 6

edly in membranes derived from PHA-stimulated lymphocytes. This contrasts with the observations on liver cells as reported by EIBL et al. [75], for reasons which have been partly discussed above.

Summarizing our results, we conclude that: 1) the lysolecithinacyltransferases are very differently distributed in the cells of various tissues. This is also true for the capacity of acylating isomeric lysolecithins. 2) In all cells so far tested, lecithin synthesis via lysolecithin acylation far exceeds its *de novo* formation. In erythrocytes this is the only active pathway and its failure seems to be involved in the elimination of aged cells. 3) For reasons which remain to be defined, in lymphocytes changes of these pathways occur very soon after a differentiative stimulus.

Acknowledgements

Our results were obtained in collaboration with H. FISCHER, W. IMM, A. KOHLSCHÜTTER, P.-G. MUNDER, H. PETERS, K. RESCH, Max-Planck-Institut, Freiburg/Br., and G. GERISCH, Friedrich-Miescher-Laboratorium, Tübingen. We thank Dr. D. F. H. WALLACH, Boston, for helpful discussions and advice.

References

1. WALLACH, D. F. H., GORDON, A.: Fed. Proc. **27**, 1263 (1968).
2. — — In: Regulatory functions of biological membranes, p. 87. (JÄRNEFELT, J., Ed.). Amsterdam: Elsevier 1968.
3. — — In: Protides of biological fluids (PEETERS, H., Ed.). Proc. 15th Coll., Brugge 1967. Amsterdam: Elsevier 1968.
4. BENSON, A. A.: In: Membrane models and the formation of biological membranes, p. 190. (BOLIS, L., PETHICA, B. A., Eds.). North Holland Publishing Co. 1968.
5. WALLACH, D. F. H., GORDON, A.: In: Membrane models and the formation of biological membranes, p. 293. (BOLIS, L., PETHICA, B. A., Eds.). North Holland Publishing Co. 1968.
6. LENARD, J., SINGER, S. J.: Proc. nat. Acad. Sci. (Wash.) **56**, 1828 (1966).
7. WALLACH, D. F. H., ZAHLER, H. P.: Proc. nat. Acad. Sci. (Wash.) **56**, 1552 (1966).
8. LENARD, J., SINGER, S. J.: Science **159**, 738 (1968).
9. SAUNDERS, L.: Biochim. biophys. Acta (Amst.) **125**, 70 (1966).
10. COLLIER, H. B.: J. gen. Physiol. **35**, 617 (1952).
11. MUNDER, P. G., FERBER, E., FISCHER, H.: Z. Naturforsch. **20b**, 1048 (1965).
12. HOWELL, J. I., LUCY, J. A.: Febs Letters **4**, 147 (1969).
13. POOLE, A. R., HOWELL, J. I., LUCY, J. A.: Nature (Lond.) **227**, 810 (1970).
14. BURDZY, K., MUNDER, P. G., FISCHER, H., WESTPHAL, O.: Z. Naturforsch. **19b**, 1118 (1964).
15. RESCH, K., FERBER, E., ODENTHAL, J., FISCHER, H.: Europ. J. Immunol. **3**, 162 (1971).
16. TATTRIE, N. H.: J. Lipid Res. **1**, 60 (1959).
17. VAN GOLDE, L. M. G., ZWAAL, R. F. A., VAN DEENEN, L. L. M.: Koninkl. Nederl. Akad. Wetensch. **68**, 255 (1965).
18. DEBUCH, H.: Z. physiol. Chem. **304**, 109 (1956).
19. HANAHAN, D. J.: Lipide chemistry. New York: John Wiley and Sons, Inc. 1960.
20. RHODES, D. N.: Biochem. J. **68**, 380 (1958).

21. HANAHAN, D. J.: J. biol. Chem. **211**, 313 (1954).
22. MUDD, J. B., VAN GOLDE, L. M. G., VAN DEENEN, L. L. M.: Biochim. biophys. Acta (Amst.) **176**, 547 (1969).
23. LANDS, W. E. M.: J. biol. Chem. **235**, 2233 (1960).
24. — MERKL, I.: J. biol. Chem. **238**, 898 (1963).
25. — HART, P.: J. biol. Chem. **240**, 1905 (1965).
26. VAN DEN BOSCH, H., VAN GOLDE, L. M. G., EIBL, H., VAN DEENEN, L. L. M.: Biochim. biophys. Acta (Amst.) **144**, 613 (1967).
27. STOFFEL, W., DE TOMÁS, M. E., SCHIEFER, H.-G.: Z. physiol. Chem. **348**, 882 (1967).
28. BRANDT, A. E., LANDS, W. E. M.: Biochim. biophys. Acta (Amst.) **144**, 605 (1967).
29. SARZALA, M. G., VAN GOLDE, L. M. G., DE KRUYFF, B., VAN DEENEN, L. L. M.: Biochim. biophys. Acta (Amst.) **202**, 106 (1970).
30. VAN DEN BOSCH, H., VAN GOLDE, L. M. G., SLOTBOOM, A. J., VAN DEENEN, L. L. M.: Biochim. biophys. Acta (Amst.) **152**, 694 (1968).
31. — VAN DEENEN, L. L. M.: Biochim. biophys. Acta (Amst.) **106**, 326 (1965).
32. SCHERPHOF, G. L., WAITE, B. M., VAN DEENEN, L. L. M.: Biochim. biophys. Acta (Amst.) **125**, 406 (1966).
33. POSSMAYER, F., SCHERPHOF, G. L., DUBBELMAN, T. M. A. R., VAN GOLDE L. M. G., VAN DEENEN, L. L. M.: Biochim. biophys. Acta (Amst.) **176**, 95 (1969).
34. VAN GOLDE, L. M. G., SCHERPHOF, G. L., VAN DEENEN, L. L. M.: Biochim. biophys. Acta (Amst.) **176**, 635 (1969).
35. HILL, E. E., LANDS, W. E. M.: Biochim. biophys. Acta (Amst.) **152**, 645 (1968).
36. STOFFEL, W., SCHIEFER, H.-G., WOLF, G. D.: Z. physiol. Chem. **347**, 102 (1966).
37. HILL, E. E., HUSBANDS, D. R., LANDS, W. E. M.: J. biol. Chem. **243**, 4440 (1968).
38. HUSBANDS, D. R., LANDS, W. E. M.: Biochim. biophys. Acta (Amst.) **202**, 129 (1970).
39. HAJRA, A. K., AGRANOFF, B. W.: J. biol. Chem. **243**, 1617 (1968).
40. — J. biol. Chem. **243**, 3458 (1968).
41. — AGRANOFF, B. W.: J. biol. Chem. **243**, 3542 (1968).
42. — Biochem. biophys. Res. Commun. **33**, 929 (1968).
43. — AGRANOFF, B. W.: Fed. Proc. **28**, 595 (1969).
44. SUNDLER, R., ÅKESSON, B.: Biochim. biophys. Acta (Amst.) **218**, 89 (1970).
45. JOHNSTON, J. M., PAULTAUF, F., SCHILLER, C. M., SCHULTZ, L. D.: Biochim. biophys. Acta (Amst.) **218**, 124 (1970).
46. RAJU, P. K., REISER, R.: Biochim. biophys. Acta (Amst.) **202**, 212 (1970).
47. KÖGEL, F., DE GIER, J., MULDER, E., VAN DEENEN, L. L. M.: Biochim. biophys. Acta (Amst.) **43**, 95 (1960).
48. MULDER, E., VAN DEENEN, L. L. M.: Biochim. biophys. Acta (Amst.) **106**, 348 (1965).

49. WAKU, K., LANDS, W. E. M.: J. Lipid Res. **9**, 12 (1968).
50. MUNDER, P. G., FERBER, E., FISCHER, H.: Z. Naturforsch. **20b**, 1048 (1965).
51. LÖHR, G. W., WALLER, H. D., KARGES, O., SCHLEGEL, B., MÜLLER, A. A.: Klin. Wschr. **36**, 1008 (1958).
52. — — Klin. Wschr. **37**, 833 (1959).
53. — — Folia haemat. (Lpz.) **78**, 385 (1962).
54. SHOJANIA, A. M., ISRAELS, L. G., ZIPURSKY, A.: J. Lab. clin. Med. **71**, 41 (1968).
55. DANON, D.: Proc. 11. Congr. Int. Soc. Blood Transf. 1966; Bibl. haemat. (Basel) **29**, 178 (1968).
56. — In: Permeability and function of biological membranes, p. 57. (BOLIS, L., KATCHALSKY, A., KEYNES, R. D., LOEWENSTEIN, W. R., PETHICA, B. A., Eds.). Amsterdam: North Holland Publishing Co. 1970.
57. FERBER, E., MUNDER, P. G., KOHLSCHÜTTER, A., FISCHER, H.: Europ. J. Biochem. **5**, 395 (1968).
58. KOHLSCHÜTTER, A., FERBER, E., MUNDER, P. G., FISCHER, H.: Folia haemat. (Lpz.) **90**, 233 (1968).
59. FERBER, E., KRÜGER, J., MUNDER, P. G., KOHLSCHÜTTER, A., FISCHER, H.: In: Stoffwechsel und Membranpermeabilität von Erythrocyten und Thrombocyten, S. 393. I. Internat. Symp. Wien 1968. Stuttgart: Thieme 1968.
60. — KOHLSCHÜTTER, A., MUNDER, P. G., FISCHER, H.: In: Modern problems of blood preservation, p. 14. (SPIELMANN, W., SEIDL, S., Eds.). Stuttgart: G. Fischer 1970.
61. — MUNDER, P. G., KOHLSCHÜTTER, A., FISCHER, H.: Z. physiol. Chem. **349**, 5 (1968).
62. MUNDER, P. G., MODOLELL, M., FERBER, E., FISCHER, H.: Biochem. Z. **344**, 310 (1966).
63. BORUN, E. R., FIGUEROA, W. G., PERRY, S. M.: J. clin. Invest. **36**, 676 (1957).
64. RIGAS, D. A., KOLER, R. D.: J. Lab. clin. Med. **58**, 242 (1961).
65. FERBER, E., MUNDER, P. G., KOHLSCHÜTTER, A., FISCHER, H.: Folia haemat. (Lpz.) **90**, 224 (1968).
66. FISCHER, H., FERBER, E., HAUPT, I., KOHLSCHÜTTER, A., MODOLELL, M., MUNDER, P. G., SONAK, R.: In: Protides of the biological fluids **15**, 175 (1967), Proceedings of the 15th Coll. Bruges 1967. Elsevier Publishing Comp.
67. GERISCH, G., MALCHOW, D., WILHELMS, H., LÜDERITZ, O.: Europ. J. Biochem. **9**, 229 (1969).
68. BEUG, H., GERISCH, G., KEMPFF, S., RIEDEL, V., CREMER, G.: Exp. Cell Res. **63**, 147 (1970).
69. GERISCH, G.: 64. Tagung der Deutschen Zoologischen Gesellschaft. Stuttgart: G. Fischer 1970.
70. FERBER, E., MUNDER, P. G., FISCHER, H., GERISCH, G.: Europ. J. Biochem. **14**, 253 (1970).
71. PETERS, J. H., HAUSEN, P.: Europ. J. Biochem. **19**, 502 (1971).

72. PETERS, J. H., HAUSEN, P.: Europ. J. Biochem. **19**, 509 (1971).
73. WALLACH, D. F. H., KAMAT, V. B.: Proc. nat. Acad. Sci. (Wash.) **52**, 721 (1964).
74. — — Meth. Enzymol. **8**, 164 (1966).
75. EIBL, H., HILL, E. E., LANDS, W. E. M.: Europ. J. Biochem. **9**, 250 (1969).
76. WOOD, R.: Arch. Biochem. **141**, 174 (1970).
77. HENNING, R., KAULEN, H. D., STOFFEL, W.: Z. physiol. Chem. **351**, 1191 (1970).

Structure and Function of Hydrocarbon Chains in Bacterial Phospholipids

Peter Overath, Hans-Ulrich Schairer,
Frank F. Hill, and Ingrid Lamnek-Hirsch

Institut für Genetik der Universität, D-5000 Köln

With 2 Figures

1. The System

As with all membranes in living organisms, investigations into the properties of the bacterial cell membrane are hindered by at least three major practical difficulties. Firstly, many interesting membrane functions, in particular transport processes, can only be studied when the membrane itself is intact and the two compartments, the interior of the cell and the outside medium, are physically separated from each other. Secondly, the disruption of the cell followed by centrifugation leads to an ill-defined "membrane" fraction containing lipids as well as many water-insoluble proteins, which greatly resist purification by the methods available today. In a Gram-negative bacterium like *E. coli*, the physical characterization of this "membrane" fraction is rendered even more complicated by the presence of fragments of the surrounding cell wall. Finally, even if the purification of a transport protein, for example, is achieved, the study of its reincorporation into a membrane so as to restore its original function will only be possible after the development of new experimental techniques.

The difficulties listed above can be circumvented if the membrane is left intact but altered in a specific way so that different physiological states of the membrane may be compared. The bacterium *E. coli* is especially suitable for such a study since it is amenable to both physiological and genetic manipulation. These approaches have been applied in our experiments in respect to two components of the membrane: (1) the hydrocarbon chains of the phospholipids and (2) the transport system for β-galactosides.

With regard to the latter, it is well established that the y-gene product of the lactose operon in the membrane is required for the transport of β-galactosides. The second important feature of the system for this particular study is that the *de novo* synthesis of the y-gene product can be induced, or repressed, at will by addition, or removal, of specific low molecular weight inducers [1, 2]. Thus, the incorporation of a specific membrane protein into the membrane can be controlled.

The fatty acid chains of the phospholipids have only recently become amenable to manipulation in two micro-organisms, namely *E. coli* [3—8] and *Mycoplasma laidlawii* [9—11]. In *E. coli* this was made possible by the isolation of mutants which require unsaturated fatty acids for growth by GILBERT and VAGELOS [12]. This organism normally contains saturated fatty acids, mainly hexadecanoate, and monounsaturated fatty acids (*cis*-Δ^{11}-octadecenoate, *cis*-Δ^9-hexadecenoate). These fatty acids are synthesized by a well-established pathway as their acyl-carrier-protein derivatives [13]. The acyl residues are then transferred to glycerolphosphate and the phosphatidic acid so formed is converted into the typical membrane phospholipids of *E. coli*, phosphatidylethanolamine, phosphatidylglycerol and diphosphatidyl-glycerol [14—17]. At the lipid stage, the unsaturated fatty acid chains may, in part, be converted to their respective cyclopronane derivatives [18, 19].

Mutants auxotrophic for unsaturated fatty acids have lesions in the *fabA* or *fabB* gene [20]. To date, only the *fabA* gene product, a β-hydroxydecanoyl-thioester dehydrase, is known: this enzyme introduces a *cis*-double bond into the hydrocarbon chain. Fatty acids, added as supplements to the medium, enter the *E. coli* cell by a partially characterized transport system [21] which is specific for fatty acids with more than eight carbon atoms. The uptake of fatty acids is presumably coupled to their activation by an acyl-CoA synthetase [22]. The resulting acyl-CoA derivatives can then either be used for lipid synthesis by transfer of the acyl-residue to α-glycerol-phosphate [23, 17], or they are degraded *via* the inducible enzymes of β-oxidation [21, 22, 24, 25]. Oxidation of fatty acids can be prevented by the introduction of a second mutation into the fatty acid requiring strains, which makes the cells unable to degrade fatty acids [26, 7]. The resulting double mutant incorporates externally added fatty acids in an unaltered form into the membrane phospholipids. It will be shown that, in such a strain, the hydrocarbon chain composition of the membrane lipids can be altered by varying the fatty acids added to the medium.

2. The Change in the Fatty Acid Composition

The occurrence of mutants of *E. coli* requiring unsaturated fatty acids for growth showed for the first time that these compounds are indispensable components of this organism. Since the synthesis of saturated fatty acids is unaffected in these mutants, it appears that an *E. coli* cell containing only saturated fatty acids is not viable. This conclusion is supported by two observations: (1) all straight-chain saturated fatty acids [from 8—18] cannot replace the requirement for an unsaturated fatty acid. (2) If a growing culture of a *fab*-mutant is deprived of the required unsaturated fatty acid, growth and cell division stop after about one generation and, depending on the experimental conditions, the cells begin to lyse [4, 27].

A systematic study was started in several laboratories to find out whether the requirement for a fatty acid containing a *cis*-double bond could be satisfied by fatty acids carrying various other functional groups in the hydrocarbon chain. Table 1 lists the results of these efforts [3, 4, 6, 7, 28]. *cis*-Monoenoic acids with even-number carbon atoms from 14 to 20 can support growth of the mutants. Moreover, a considerable variation in the position of the double bond may be tolerated for growth. A number of fatty acids containing 18 and 20 carbon atoms and several *cis*-double bonds have been shown to serve as growth factors. A most surprising observation is that fatty acids with a *trans*-double bond may substitute for the normally occurring *cis*-monoenoic acids, as the lipids with saturated and/or *trans*-ethylenic hydrocarbon chains have quite different physical properties compared with lipids containing hydrocarbon chains with *cis*-double bonds (see below). Finally, it is possible to use fatty acids with a cyclopropane ring, a triple bond or a bromine substitution in the hydrocarbon chain as growth factors.

Among the fatty acids which do not support growth of the mutant are derivatives with a) *cis*-double bonds, e.g. a mixture of cis-Δ^2 and Δ^3-decenoate, cis-Δ^5-$C_{20:1}$ or cis-Δ^{13}-$C_{22:1}$, b) methyl-branches near the end of the chain, e.g. 14-methylhexadecanoic acid, or c) hydroxyl-groups, e.g. D(-)9-hydroxyoctadecanoic acid [3]. The lack of growth of the mutant with these supplements may be due to a deficiency in transport, activation or transacylation of these fatty acids rather than the unsuitability of their properties for membrane lipid function. Transport and activation of fatty acid containing more than eight carbon atoms is possible [21, 22], but an extensive study of the specificity of these processes has not yet

Table 1. *Fatty acids supporting growth of fab-auxotrophs*

Functional group(s)	examples
one cis-double bond	cis-Δ^5-$C_{14:1}$; cis-Δ^9-$C_{14:1}$; cis-Δ^6-$C_{18:1}$; cis-Δ^9-$C_{18:1}$; cis-Δ^{11}-$C_{18:1}$; cis-Δ^{11}-$C_{20:1}$;
several cis-double bonds	cis, cis-$\Delta^{9,12}$-$C_{18:2}$; cis, cis, cis-$\Delta^{9,12,15}$-$C_{18:3}$; cis, cis-$\Delta^{11,14}$-$C_{20:2}$; cis, cis, cis-$\Delta^{11,14,17}$-$C_{20:3}$;
trans-double bond	trans-Δ^9-$C_{16:1}$; trans-Δ^9-$C_{18:1}$; trans-Δ^{11}-$C_{18:1}$
cyclopropane ring	D,L-cis-9,10-methylene-octadecanoic acid
triple bond	octadec-9,10-ynoic acid
bromine	mixture of 9- and 10-bromo-stearic acid

The number before the colon gives the number of carbon atoms and the number after the colon gives the number of double bonds; superscript to Δ gives position of ethylenic bond. For references see: [3, 4, 6, 7, 28].

been performed. The same is true for the transfer of the acyl-groups from CoA to α-glycerolphosphate [17, 23]. However, an examination of the fatty acids which do support growth indicates that the specificity of transport, activation and acyl-transfer is rather broad. This is also supported by the fact that the wild type can grow on D(-)9-hydroxyoctadecanoic, 2-hydroxy-myristic or 12-hydroxy-cis-Δ^9-octadecenoic acids as sole carbon source (H. U. SCHAIRER and P. OVERATH, unpublished results). In summary, the data so far suggest the following structural requirements of the hydrocarbon chain structure in *E. coli* membrane lipids for proper function: (1) the hydrocarbon chain must have 14 to 20 carbon atoms; (2) there has to be a perturbation in the chain which results in an

Table 2. Fatty acid composition (weight %) of phosphatidylethanolamine from E. coli mutant grown with various fatty acids

	Fatty acid supplied to the growth medium				
	$cis\text{-}\Delta^9\text{-}C_{18:1}$	$trans\text{-}\Delta^9\text{-}C_{18:1}$	$trans\text{-}\Delta^9\text{-}C_{16:1}$	C_{19}^{Δ}	$cis, cis, cis\text{-}\Delta^{7,12,15}\text{-}C_{18:3}$
$C_{16:0}$	22	6	17	27	50
$C_{14:0}$	6	11	7	17	3
$C_{12:0}$	—	2	1	1	—
$cis\text{-}\Delta^9\text{-}C_{18:1}$	58	—	—	—	—
C_{19}^{Δ}	14	—	—	54	—
$trans\text{-}\Delta^9\text{-}C_{18:1}$	—	81	—	—	—
$trans\text{-}\Delta^9\text{-}C_{16:1}$	—	—	73	—	—
$cis, cis, cis\text{-}\Delta^{9,12,15}\text{-}C_{18:3}$	—	—	—	—	46
Unidentified	—	< 1	2	2	1
$C_{16:0}$	3.7	0.55	2.4	1.6	17
$\overline{C_{14:0}}$					
Ratio of fatty acids[a]	0.39	0.23	0.34	0.83	1.2

[a] Sum of the saturated fatty acids divided by the sum of the unsaturated and cyclopropane derivatives. Data taken from [26, 29]. C_{19}^{Δ} refers to D,L-cis-9,10-methylene-octadecanoic acid.

increase in the rotational volume of the chain. This requirement will become clearer from the functional analysis presented below.

The extent of incorporation of the fatty acids listed in Table 1 is quite variable and depends both on the fatty acid and the particular mutant employed [3, 4, 6, 7]. Table 2 lists some of the data obtained in our laboratory with a mutant (K 1059) unable to synthesize or degrade unsaturated fatty acids. As can be seen from the ratio given at the bottom of the table, the bulky linolenic acid is incorporated to a lesser extent than the other fatty acids, especially those with a *trans*-double bond in the chain. This effect may indicate a control mechanism operating at the trans-acylation step of lipid synthesis [30] which tends to optimize the hydrocarbon chain composition so that it is adequate for growth. The change in the synthesis of saturated fatty acids, as shown by the ratio of $C_{16:0}$ to $C_{14:0}$ suggests a similar attempt by the cell to achieve optimal hydrocarbon chain fluidity *via* control of the enzymatic machinery for fatty acid formation.

3. Conditional Lethal Growth on Fatty Acids

Although the fatty acids listed in Tables 1 and 2 do support growth of the mutants, and therefore the lipids containing these fatty acids seem to be adequate for growth, this may not be true for all the conditions which support growth of the wild type. Thus, it can be shown that the range of temperature over which the wild type, or the mutants in the presence of *cis*-Δ^9-$C_{18:1}$, can grow is diminished if the mutants are supplied with *trans*-unsaturated fatty acids or with *cis*-polyunsaturated fatty acids. Table 3 lists the temperature range for growth of the wild type *E. coli* [31] and the mutant K 1059 [26]. With *cis*-Δ^9-$C_{18:1}$ as a supplement, the growth rate decreased sharply at 10 °C and 44 °C indicating similar extremes as the wild type. The polyunsaturated linolenic acid reduces the maximum growth temperature to 40 °C, while the minimum growth temperature is the same as in the wild type. On the other hand, growth of the mutant on *trans*-Δ^9-$C_{18:1}$ allows the same growth maximum as the wild type while raising the minimum growth temperature to 37 °C. The lethal effect of a downward shift in temperature has been independently observed by ESFAHANI et al. [7] with *trans*-Δ^{11}-$C_{18:1}$ and *cis*-Δ^6-$C_{18:1}$.

The following conclusions can be drawn from these experiments. (1) In the wild type neither the maximum nor the minimum temperature for growth is a function of the fatty acid composition of the phospholipids. If this were the case, one would expect a rise in the maximum growth temperature when $trans$-Δ^9-$C_{18:1}$ is supplied and a decrease in the minimum growth temperature when linolenic acid is supplied. (2) The fatty acid composition may give rise to a decrease in the maximum temperature of growth when a highly unsaturated fatty acid is supplied. This suggests that there is an upper limit of fluidity of the lipid phase compatible with

Table 3. *Temperature range of growth of E. coli*

Strain	fatty acid added to the growth medium	temperature range of growth
K 12 (wild type)[a]	—	8—46 °C
K 1059[b]	cis-Δ^9-$C_{18:1}$	10—44 °C
K 1059[b]	cis, cis, cis-$\Delta^{9,12,15}$-$C_{18:3}$	10—40 °C
K 1059[b]	$trans$-Δ^9-$C_{18:1}$	37—45 °C

[a] INGRAHAM, J. L. [31].
[b] OVERATH, P., SCHAIRER, H. U., and STOFFEL, W. [26].

growth. Finally, the minimum temperature for growth can be raised if *trans*-unsaturated fatty acids are incorporated into the membrane lipids.

4. Thermotropic Phase Transitions of Membrane Lipids *in vivo*

The increase in the minimum temperature for growth from 8 °C in the wild type to 37 °C for the mutant supplied with $trans$-Δ^9-$C_{18:1}$ as growth factor, suggested that there may be a change in the arrangement of the hydrocarbon chains at 37 °C which impairs various functions associated with the membrane, resulting in ultimate cessation of growth. Extensive investigations with a variety of physical techniques, especially in the laboratories of D. CHAPMAN in England and V. LUZZATI in France [32], have established that pure lipids can suffer reversible, endothermic phase transitions from a crystalline to a liquid-crystalline state. STEIM et al. [33],

MELCHIOR et al. [34] and ENGELMAN [35, 36] have shown by differential scanning calorimetry and by X-ray diffraction that thermotropic phase changes occur in membranes of *Mycoplasma laidlawii*. It is generally agreed that these phase transitions reflect the conversion of an ordered to a disordered, mobile state of the hydrocarbon chains in the lipid phase. The most direct evidence for the molecular motion of the hydrocarbon chains in phospholipid dispersions and in several biological membranes has been obtained by MCCONNELL and co-workers [37, 38] by analysis of the electron paramagnetic resonance spectra of spin-labeled phospholipids dissolved in the membrane preparations.

The detection of phase transition in *E. coli* was made possible by the use of the mutants described in the preceding paragraphs: 1) The temperature characteristic of β-galactoside transport in *E. coli*-cells, grown in the presence of various fatty acids, shows a biphasic behavior with a sharp transition point. 2) If the transition points measured *in vivo* reflect a change in the arrangement of hydrocarbon chains, it should be possible to discover these phase transitions with the isolated lipids *in vitro*. The results of such experiments are summarized in Table 4. The first parameter listed in Table 4 is the efflux of thiomethyl-β-galactoside, a carrier-mediated transport process which is independent of metabolic energy [40—42]. Above the transition points the temperature characteristic of the efflux, μ, is 17 to 20 kcal/mole, below the transition points values of 42 to 46 kcal/mole are found. The simplest explanation of this effect is that, below the transition, the mobility of the carrier(s) become(s) impaired as a result of the rearrangement of the hydrocarbon chains in the lipid phase. A similar parameter, the *in vivo* hydrolysis of o-nitrophenyl-β-galactoside, which is known to be dependent on the y-gene product of the *lac*-operon [1, 43], was investigated by WILSON, ROSE and FOX [5]. Table 4 lists two transition points of the temperature characteristic of this transport process.

The main phospholipid of *E. coli*, phosphatidylethanolamine, was isolated from the cells, and monolayers were analysed by W. STOFFEL [26] by the Langmuir-technique at an air-water interface. Above the transition points listed in Table 4 (third line) the lipid films were, at all pressures, in the liquid-expanded state which is characterized by mobile hydrocarbon chains. At the temperatures

Table 4. *Correlation of in vivo and in vitro transition points*

	Fatty acid incorporated into phospholipid				References
	cis-Δ^9-$C_{18:1}$	$trans$-Δ^9-$C_{16:1}$	$trans$-Δ^9-$C_{18:1}$	cis, cis, cis-$\Delta^{9,12,15}$-$C_{18:3}$	
in vivo					
efflux of ^{14}C-thio-methyl-β-galactoside	15 °C	—	38 °C	6 °C	[26]
o-nitrophenyl-β-galactoside transport	15 °C	32.6 °C	—	—	[5, 29]
in vitro					
appearance of phase transitions in lipid[a] monolayers	15 °C	—	41 °C	4 °C	[26]
beginning of phase transitions of lipid[a]/H_2O dispersions (ANS-binding-technique)	13 °C	29.5 °C	41 °C	—	[39]

[a] The data refer to the isolated phosphatidylethanolamines with the fatty acid composition listed in Table 2.

listed, condensed phases appeared at a molecular area of approx. 50 Å²/molecule and a film pressure of approx. 50 dyn/cm. The temperatures at which these phase transitions occurred were strongly dependent on the hydrocarbon chain composition of the lipids and they correlated with the temperatures for the transition of the transport measurements.

Although the comparison of the *in vivo* and *in vitro* experiments suggested that a transition from a "liquid" to a condensed state occurred in the lipid phase of the membrane, it was not completely clear if this was really the liquid-crystalline ↔ crystalline transition observed with thermo-analytical methods [44]. Recent experiments in collaboration with H. TRÄUBLE [39] indicate that there is indeed such a relationship (Table 4, last line). TRÄUBLE [45] has been able to measure phase transitions in lipid dispersions or in aqueous solutions of monolayer lipid vessicles by an optical method with 1-anilino-8-naphthalene-sulfonate (ANS) as fluorescent indicator. The temperatures of these phase transitions correlate well with those obtained by calorimetric methods. Since the phase transitions observed with the ANS-binding technique (Table 4, line 4) begin at approximately the same temperature at which condensed phases appear in monolayers, it seems that the transition points tabulated in Table 4 do indeed reveal the liquid-crystalline ↔ crystalline phase change of the phospholipids.

5. The Random Growth of the Lipid Phase in the Membrane

The transition point in the temperature characteristic of a transport parameter specifies the hydrocarbon chain composition of the membrane lipids in the surroundings of the transport protein. Furthermore, as mentioned above, the synthesis of the β-galactoside transport protein can be induced by the addition of the inducer isopropyl-β-galactoside [1]. This technique can be used to investigate the biogenesis of the lipid phase of the bacterial membrane.

Table 5 lists some of the models which may be suggested for membrane growth. The enzymes involved in lipid synthesis are themselves firmly membrane-bound [14—17]. It is therefore conceivable that the lipid molecules are formed in the membrane. At the molecular level, newly synthesized lipid molecules may either be deposited in patches (A_1) or in a random way (A_2). At the cellular level, these possibilities can lead to at least three possible ways for the arrangement of the lipid phase (C_1—C_3). In the present context these topological considerations should be considered mainly in terms of the hydrocarbon chains of the phospholipids and not in terms of the arrangement of the polar heads in the membrane, which is currently unknown. Thus, a membrane protein may associate with lipid molecules containing specific polar heads, but there

Table 5. *Model considerations for the biogenesis of the bacterial membrane*

	Membrane growth at the molecular level	cellular level
growth of the lipid phase	A_1) in lipid patches at or near the enzymes of lipid synthesis A_2) in a random fashion involving lateral diffusion of lipid molecules	C_1) in a zonal fashion C_2) in patches at many sites C_3) as A_2
incorporation of a *de novo* synthesized protein	B_1) into "new" lipid patches B_2) into "new" and "old" lipid patches B-) into a random lipid phase	D_1) in a zonal fashion, *e.g.* in the middle of the cell D_2) at random or specifically, at many sites

may be an exchange with other molecules carrying the same polar groups but different hydrocarbon chains.

The site of synthesis of a newly formed protein at a ribosome is assumed to be independent of the site of incorporation of that protein into the membrane. At the molecular level, a protein may therefore be incorporated into newly formed lipid patches (B_1), into new and old lipid patches (B_2) or into a random lipid phase (B_3). Finally, at the cellular level, proteins may be incorporated in a specific growth zone (D_1) or at many sites in the cell membrane (D_2). Growth of the membrane at a growth zone in the middle of the cell has been postulated by JACOB et al. [46].

The growth of the lipid phase in relation to the *de novo* synthesis of the β-galactoside transport system was first investigated by Fox et al. [47—49]. These authors concluded from their experiments that the newly synthesized transport protein is incorporated together with newly formed lipids into the membrane (A_1 and B_1 in Table 5). However, various other results disagree with this conclusion [50, 51].

Evidence for randomization of lipid molecules (A_2 and B_3 in Table 5) in the membrane was obtained from experiments of the type shown in Fig. 1. *E. coli* strain K 1062 [29], a mutant unable to synthesize or degrade unsaturated fatty acids, was grown with oleate (*cis*-Δ^9-$C_{18:1}$) as supplement (Fig. 1a). The cells were then transferred to a medium containing palmitelaidate (*trans*-Δ^9-$C_{16:1}$) instead of oleate as growth factor. At various times after this change in the medium, the transport system for β-galactosides was induced in separate samples by addition of isopropyl-β-galactoside. Addition and removal of inducer is indicated by arrows. Fig. 1b indicates a similar experiment involving a shift from palmitelaidate to oleate. At the end of the induction periods, the cells were harvested and the temperature characteristics of o-nitrophenylgalactoside (ONPG) hydrolysis was measured. In parts c and d of Fig. 1, the cells were induced in the presence of the first fatty acid and further growth was then allowed in the presence of the second fatty acid but in the absence of inducer. The temperature characteristics of ONPG-hydrolysis was subsequently determined after various growth periods in the second medium.

Fig. 2 a—d shows the results of these experiments. The reciprocal absolute temperature is plotted against the logarithm of the relative rate of ONPG-hydrolysis. The designation a to d corresponds to that in Fig. 1, and the percentage of growth during (a, b) or after (c, d) induction is shown at the top of the curves. The curves for cells induced in the presence of oleate (left-hand curve of Fig. 1a) or palmitelaidate (left-hand curve Fig. 1b) alone give transition points of 14.9 °C and 33.4 °C, respectively (compare Table 4). It is evident from Fig. 2 that there is a continuous rise (a, c) or fall (b, d) of the transition points between the extremes 14.9 °C and 33.4 °C, which corresponds to the extent of growth in the medium after the shift. A plot of the mole fraction of oleate incorporated into the lipids to the sum of all fatty acids *versus* the transition points in

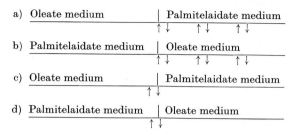

Fig. 1. Design of the experiments shown in Fig. 2. The horizontal line indicates the time scale and the vertical line indicates the medium shift. The arrows refer to the addition (↑) or removal (↓) of the inducer of the *lac*-operon

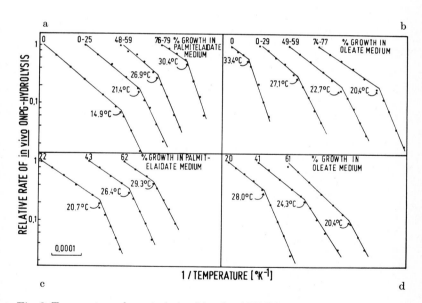

Fig. 2. Temperature characteristic of *in vivo* ONPG-hydrolysis after a fatty acid shift. For details see text and [29]

Fig. 2a gives a straight line between the extremes [29]. The same result is obtained if the transport protein is incorporated into the membrane while the fatty acid composition is changed, or if it is first incorporated and a change in the lipid composition follows. Under all conditions the temperature characteristic is determined by the average fatty acid composition of the membrane lipids. Our data, therefore, support possibilities A_2 and B_3 listed in Table 5. Since an enzymatic mechanism for randomization has been excluded [29], it is suggested that lateral diffusion of lipid molecules leads to mixing of the lipid phase. This assumption is supported by measurements (H. TRÄUBLE and E. SACKMAN, personal communication) which demonstrate a rapid diffusion (average travel distance 10000 Å/sec) of a spin-labeled androstane in synthetic lipid bilayers.

6. A Lipid Bilayer in the Bacterial Membrane

The thermotropic phase changes observed in *E. coli* demonstrate that there is lipid-lipid interaction in the membrane. This idea is clearly in accordance with a bilayer structure for the arrangement of the lipids in the membrane. The recent analysis of several biological membranes, including the cell membrane of *E. coli*, by X-ray diffraction provides conclusive evidence for a lipid bilayer structure [36, 52, 53]. The impairment in the rates of carrier-mediated transport processes below the transition point suggests that transport proteins require a mobile state of the hydrocarbon chains in the lipid phase. Mobility in the lipid phase may be a universal requirement for proper membrane function. The lateral diffusion of lipid molecules in the membrane may simply be another aspect of the bilayer structure. Randomization of lipid molecules, however, does not necessarily imply a random incorporation of proteins or protein complexes into the membrane.

This work was supported through the SFB 74 "Molekularbiologie der Zelle" from the Deutsche Forschungsgemeinschaft.

References

1. RICKENBERG, H. V., COHEN, G. N., BUTTIN, G., MONOD, J.: Ann. Inst. Pasteur **91**, 829 (1956).
2. KENNEDY, E. P.: In: The lactose operon. (BECKWITH, J. R., ZIPSER, D., Eds.). Cold Spr. Harb. Labor., **1970**, 49.
3. SILBERT, D. F., RUCH, F., VAGELOS, P. R.: J. Bact. **95**, 1658 (1968).
4. SCHAIRER, H. U., OVERATH, P.: J. molec. Biol. **44**, 209 (1969).
5. WILSON, G., ROSE, S. P., FOX, C. F.: Biochem. biophys. Res. Commun. **38**, 617 (1970).

6. ESFAHANI, M., BARNES, E. M., WAKIL, S. J.: Proc. nat. Acad. Sci. (Wash.) **64**, 1057 (1969).
7. — IONEDA, T., WAKIL, S. J.: J. biol. Chem. **246**, 50 (1971).
8. SILBERT, D. F.: Biochemistry **9**, 3631 (1970).
9. RODWELL, A.: Science **160**, 1350 (1968).
10. HENDRIKSON, C. V., PANOS, C.: Biochemistry **8**, 646 (1969).
11. MCELHANEY, R. N., TOURTELLOTTE, M. E.: Biochim. biophys. Acta (Amst.) **202**, 120 (1970).
12. SILBERT, D. F., VAGELOS, P. R.: Proc. nat. Acad. Sci. (Wash.) **58**, 1579 (1967).
13. VAGELOS, P. R.: Ann. Rev. Biochem. **33**, 139 (1964).
14. KANFER, J., KENNEDY, E. P.: J. biol. Chem. **239**, 1720 (1964).
15. AILHAUD, G. P., VAGELOS, P. R.: J. biol. Chem. **241**, 3866 (1966).
16. STANACEV, N. Z., CHANG, Y., KENNEDY, E. P.: J. biol. Chem. **242**, 3018 (1967).
17. VAN DEN BOSCH, H., VAGELOS, P. R.: Biochim. biophys. Acta (Amst.) **218**, 233 (1970)
18. LAW, J. H., ZALKIN, H., KENESHIRO, T.: Biochim. biophys. Acta (Amst.) **70**, 143 (1963)
19. ZALKIN, H., LAW, J. H., GOLDFINE, H.: J. biol. Chem. **238**, 1242 (1963).
20. CRONAN, J. E. Jr., BIRGE, C. H., VAGELOS, P. R.: J. Bact. **100**, 601 (1969).
21. KLEIN, K., STEINBERG, R., FIETHEN, B., OVERATH, P.: Europ. J. Biochem. **19**, 442 (1971).
22. OVERATH, P., PAULI, G., SCHAIRER, H. U.: Europ. J. Biochem. **7**, 559 (1969).
23. PIERINGER, R. A., BONNER, H., KUNNES, R. S.: J. biol. Chem. **242**, 2719 (1967).
24. OVERATH, P., RAUFUSS, E. M., STOFFEL, W., ECKER, W.: Biochem. biophys. Res. Commun. **29**, 28 (1967).
25. WEEKS, G., SHAPIRO, M., BURNS, R. O., WAKIL, S. J.: J. Bact. **97**, 827 (1969).
26. OVERATH, P., SCHAIRER, H. U., STOFFEL, W.: Proc. nat. Acad. Sci. (Wash.) **67**, 606 (1970).
27. HENNING, U., DENNERT, G., REHN, K., DEPPE, G.: J. Bact. **98**, 784 (1969).
28. FOX, C. F., LAW, J. H., TSUKAGOSHI, N., WILSON, G.: Proc. nat. Acad. Sci. (Wash.) **67**, 598 (1970).
29. OVERATH, P., HILL, F. F., LAMNEK, I.: Nature New Biol. (submitted for publication).
30. SINENSKY, M.: J. Bact. **106**, 449 (1971).
31. INGRAHAM, J. L.: J. Bact. **76**, 75 (1958).
32. CHAPMAN, D.: Biological membranes. Physical fact and function. New York: Academic Press 1968.
33. STEIM, J. M., TOURTELLOTTE, M. E., REINERT, J. C., MCELHANEY, R. N., RADER, R. L.: Proc. nat. Acad. Sci. (Wash.) **63**, 104 (1969).
34. MELCHIOR, D. L., MOROWITZ, H. J., STURTEVANT, J. M., TSONG, T. Y.: Biochim. biophys. Acta (Amst.) **219**, 114 (1970).

35. ENGELMAN, D. M.: J. molec. Biol. **47**, 115 (1970).
36. — J. molec. Biol. **58**, 153 (1971).
37. HUBBELL, W. L., McCONNELL, H. M.: J. Ann. Chem. Soc. **93**, 314 (1971).
38. McCONNELL, H. M.: In: The neurosciences, Second Study Program, p. 697 (1970). (SCHMITT, F. O., Ed.).
39. TRÄUBLE, H., OVERATH, P.: unpublished results.
40. KEPES, A.: In: The molecular basis of membrane function, p. 353 (TOSTESON, D. C., Ed.). Prentice-Hall: Englewood Cliffs 1969.
41. KOCH, A. L.: Biochim. biophys. Acta (Amst.) **79**, 177 (1964).
42. WINKLER, H. H., WILSON, T. H.: J. biol. Chem. **241**, 2200 (1966).
43. FOX, C. F., KENNEDY, E. P.: Proc. nat. Acad. Sci. (Wash.) **54**, 891 (1965).
44. PHILLIPS, M. C., WILLIAMS, R. M., CHAPMAN, D.: Chem. Phys. Lipids **3**, 234 (1969).
45. TRÄUBLE, H.: Naturwissenschaften **58**, 277 (1971).
46. JACOB, F., BRENNER, S., CUZIN, F.: Cold Spr. Harb. Symp. quant. Biol. **28**, 329 (1963).
47. FOX, C. F.: Proc. nat. Acad. Sci. (Wash.) **63**, 850 (1969).
48. WILSON, G., FOX, C. F.: J. molec. Biol. **55**, 49 (1971).
49. HSU, C. C., FOX, C. F.: J. Bact. **103**, 410 (1970).
50. MINDICH, L.: Proc. nat. Acad. Sci. (Wash.) **68**, 420 (1971).
51. WILLECKE, K., MINDICH, L.: J. Bact. **106**, 514 (1971).
52. WILKINS, M. H. F., BLAUROCK, A. E., ENGELMAN, D. M.: Nature New Biol. **230**, 72 (1971).
53. CASPAR, D. L. D., KIRSCHNER, D. A.: Nature New Biol. **231**, 46 (1971).

Some Aspects of the Structure and Assembly of Bacterial Membranes

LAWRENCE I. ROTHFIELD

Department of Microbiology, University of Connecticut Health Center, Farmington, CT 06032, U.S.A.

With 7 Figures

Bacteria offer many advantages in studies of biological membranes, not the least of which is the absence of the complicated internal membrane structures of eukaryotic cells. The present dis-

Table 1. *Functions located in inner and outer membranes*

Inner (plasma) membrane	Outer membrane
Active transport	Barrier function
Electron transfer reactions	Endotoxin
Lipopolysaccharide biosynthesis	Phage receptors
(?) Peptidoglycan biosynthesis	Somatic (0) antigens
(?) DNA replication	
(?) Protein synthesis	

cussion will concern the membranes of gram-negative bacteria. In these organisms the only two membrane structures are both located in the periphery of the cell. The inner, or plasma membrane, immediately surrounds the cytoplasm. It is separated from a second membrane, the outer membrane, by the so-called "periplasmic space" [1]. Both membranes show the characteristic "unit membrane" structure by electron microscopy, and both consist primarily of lipids and proteins. The two membranes serve different functions, however, as indicated in Table 1.

The inner membrane carries out the usual functions of the plasma membrane of eukaryotic cells, but also participates in functions located in the membranes of mitochondria and other intracellular organelles in higher organisms. In addition, enzymes present in the inner membrane participate in the synthesis of a variety of macromolecules including lipopolysaccharide [2] and probably peptidoglycan. The inner membrane also contains a complement of bound ribosomes to approximately 15 to 20% of the

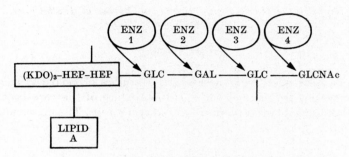

Fig. 1. Partial structure of lipopolysaccharide of *Salmonella typhimurium*. A portion of the core region of the lipopolysaccharide is shown in simplified form (for a more complete diagram see reference [15]). The site of action of four of the nucleotide sugar: lipopolysaccharide glycosyl transferase enzymes is indicated for illustrative purposes. Enz 1 and Enz 2 refer to UDP-glucose: lipopolysaccharide glucosyl transferase I and UDP-galactose: lipopolysaccharide $\alpha,3$ galactosyl transferase, respectively [15]

total cellular ribosomes but it is not known whether these membrane-bound ribosomes participate in synthesis of specific classes of proteins.

The known functions of the outer membrane are much fewer, and several of these are due to the presence of an unusual glycolipid, the characteristic lipopolysaccharide of gram-negative bacteria. This compound consists of a lipid portion (Lipid A) covalently linked to a complex phosphorylated polysaccharide (Fig. 1). Most of the lipopolysaccharide of the cells is located in the outer membrane although a small amount may also be present in the inner membrane. The lipopolysaccharide is responsible for the endotoxin activity of gram-negative bacteria and the polysaccharide portion of the molecule serves as a receptor site for several bacteriophages and contains the 0-antigen determinants which are the basis for the serological typing of gram-negative bacteria [3].

The many different functions served by these membranes are reflected by a large number of proteins in addition to a lesser variety of lipid components. Thus, despite the apparent simplicity of arrangement of membranes in these organisms, their great functional and compositional heterogeneity make it difficult to study their molecular organization or manner of assembly in detail. To avoid some of the difficulties of studying such a complex multicomponent organelle, it is helpful to view the membrane as a series of functional subunits. Each subunit is viewed as composed of a functional protein or proteins together with the lipids and other membrane components associated with the system. Individual functional subunits can then be studied by classical biochemical techniques of dissociation, purification and reconstitution. When many such systems are defined, the structure of the membrane should emerge.

Glycosyl Transferases

The difficulty with this approach has been the general inability to obtain functional membrane components in purified form. In recent years, however, as techniques of lipid and protein chemistry have improved, functional components of several membrane systems have become available for study. The present discussion will center on one such functional subunit of the inner membrane of *Salmonella typhimurium*. The protein components of this system are a series of glycosyl transferase enzymes which catalyze the sequential addition of sugar residues to the polysaccharide portion of the lipopolysaccharide molecule [4]. The site of action of several of these enzymes is schematically shown in Fig. 1. Two of the enzymes have been obtained in pure form (Enz I and Enz II in Fig. 1), catalyzing the sequential incorporation of glucose and galactose into the lipopolysaccharide [5, 6]. Reconstitution of activity of these enzymes *in vitro* requires the presence of phosphatidylethanolamine, the major membrane phospholipid in these organisms, in addition to the enzyme protein and the lipopolysaccharide acceptor. The reactions catalyzed by the two enzymes are:

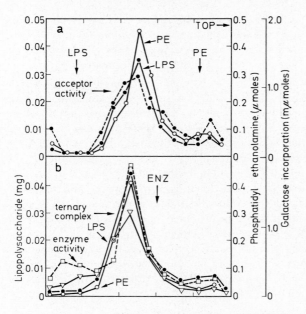

Fig. 2. Isolation of (a) lipopolysaccharide-phosphatidylethanolamine and (b) enzyme-lipopolysaccharide-phosphatidylethanolamine complexes by isopycnic centrifugation (from WEISER and ROTHFIELD [7]). (a) Mixtures of lipopolysaccharide (LPS) and phosphatidylethanolamine (PE) were applied to the gradient. (b) Galactosyl transferase (ENZ) was added prior to centrifugation. Acceptor activity was assayed by adding UDP-[^{14}C]galactose and galactosyl transferase enzyme to each gradient fraction; enzyme activity was assayed by adding excess lipopolysaccharide phosphatidylethanolamine acceptor and UDP-[^{14}C]galactose; ternary complex was assayed by adding UDP-[^{14}C]galactose. The arrows indicate the positions of each of the compounds when centrifuged under the same conditions but in the absence of the other two components

$$\text{UDP-Glc} + \text{LPS}^1 \longrightarrow \text{Glc-LPS} + \text{UDP} \qquad (1)$$
$$\text{UDP-Gal} + \text{Glc-LPS} \longrightarrow \text{Gal-Glc-LPS} + \text{UDP}.$$

The reconstituted systems appear nearly identical to the native systems within the membrane when various enzymatic parameters of the reactions are measured. Therefore, the details of the molecular

---

[1] Abbreviations — LPS, lipopolysaccharide.

organization and mechanism of reassembly of the reconstituted systems should provide information relevant to their organization and biogenesis within the cell.

Reassembly of the components *in vitro* takes place in a stepwise fashion. The first step requires the interaction of the two lipid components of the system, lipopolysaccharide and phosphatidylethanolamine. This results in formation of a multimolecular complex which has been isolated by isopycnic centrifugation in sucrose (Fig. 2a) [7]. The enzyme proteins can then be incorporated into the lipopolysaccharide-phosphatidylethanolamine complex with formation of a ternary complex which has also been isolated by isopycnic centrifugation (Fig. 2b). In a similar manner the second available enzyme, the glucosyl transferase, can also be incorporated into a common particle containing lipopolysaccharide, phosphatidylethanolamine, glucosyl transferase and galactosyl transferase.

Monolayer Studies

A technique that has proven of great value in studying the details of this reassembly process is the technique of monolayer penetration. In these experiments, a monomolecular film of phosphatidylethanolamine was first spread on the surface of an aqueous subsolution. The phospholipid molecules orient themselves in a molecular monolayer, with hydrocarbon chains directed upward into the air phase while the polar head groups face the aqueous subsolution. It was then possible to sequentially introduce lipopolysaccharide and transferase enzymes into the phospholipid monolayer and to demonstrate restoration of transferase enzyme function within the protein-lipopolysaccharide-phosphatidylethanolamine film.

To accomplish the sequential incorporation of components into a monomolecular film, a special device was used permitting movement of the monolayer from the surface of one subsolution to the surface of a fresh subsolution so that the reassembly could be accomplished in stepwise fashion. The experimental plan which demonstrated incorporation of galactosyl transferase into the monolayer is illustrated in Fig. 3. A monomolecular film of phosphatidylethanolamine was first formed and lipopolysaccharide and galactosyl transferase were sequentially introduced into the subphase.

Penetration of these molecules into the phospholipid monolayer was indicated by an increase of surface pressure (Fig. 4) and the restoration of enzyme activity in the monolayer was demonstrated by introducing UDP-[^{3}H]galactose into the subsolution and directly measuring transfer of [^{3}H]galactose into the film. As shown in

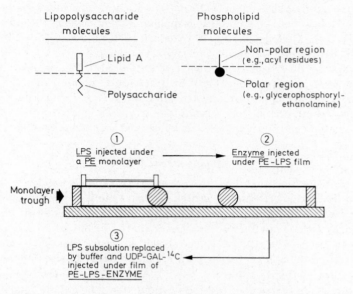

Fig. 3. Reconstitution of the galactosyl transferase system in a monomolecular film

Table 2, restoration of the transferase activity required all three components of the system.

The reconstituted system in the monolayer was similar in many respects to the native membrane when several enzymatic parameters were studied including the K_m for UDP-galactose and the turnover number of the enzyme.

Calculation of the molecular surface area of each of the components in the monolayer revealed a surface area of approximately 54 A^2 for each molecule of phosphatidylethanolamine. This is consistent with the well-established orientation of phospholipids at the air-water interface, with acyl chains directed upward and polar head groups directed downward. The calculated surface area of the lipopolysaccharide was approximately 240 A^2 per molecule. Examination of molecular models of lipopolysaccharide indicated that the conformation illustrated in Fig. 5 was fully consistent with

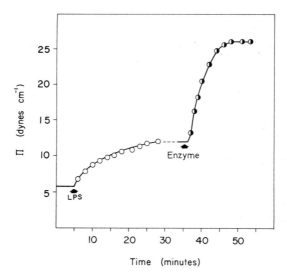

Fig. 4. Penetration of lipopolysaccharide and galactosyl transferase into a monomolecular film of phosphatidylethanolamine (from ROMEO, HINCKLEY and ROTHFIELD [9]). Lipopolysaccharide (LPS) and galactosyl transferase enzyme were sequentially injected beneath a monolayer of phosphatidylethanolamine at the times indicated by the arrows. π, surface pressure of the film

the observed results. In this conformation the lipopolysaccharide molecule takes up an area of approximately 250 A^2 in the plane of the film. The fatty acid groups of the Lipid A portion of the molecule are directed upward into the air phase while the polar polysaccharide portion is oriented downward. This conformation permits phospholipid molecules to associate closely in a side-by-side arrangement with lipopolysaccharide on the surface of the subsolution as indicated in Fig. 5b. The methylene groups of the acyl chains of

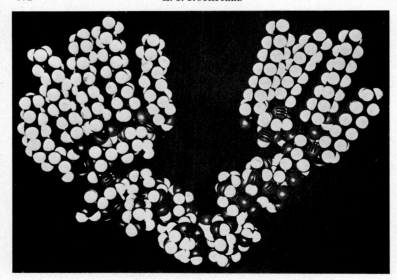

a

b

Fig. 5

the two molecules come within 2 to 3 A of each other, close enough to permit non-polar interactions of the VAN DER WAALS type to occur.

Since a mixed monolayer of this type, consisting of a two-dimensional array of lipopolysaccharide and phosphatidylethanolamine in a side-by-side configuration, is analogous to one-half of a molecular bilayer, the results suggest that the portion of the membrane containing the galactosyl transferase enzyme is probably a

Table 2. *Galactosyl transferase activity in monolayers*

Film components	Galactose incorporation (p moles/5 min)
LPS + PE + Enzyme	39
LPS + Enzyme	< 5
PE + Enzyme	< 5
None	< 5

UDP-[^{3}H]galactose was introduced beneath the film and [^{3}H]galactose incorporation into the film was measured at intervals. (From ROMEO, HINCKLEY and ROTHFIELD [9].)

mixed bimolecular leaflet as indicated in Fig. 6. The exact location of the enzyme protein has not been established. It is certain that a portion of the enzyme is located in the polar portion of the film since the protein catalyzes transfer of galactose to the terminal sugar residue of the polysaccharide chain, which is directed downward (see Fig. 5). It is not yet known whether a portion of the enzyme also penetrates into the hydrocarbon interior of the membrane.

In a manner similar to that described above for the galactosyl transferase enzyme, it also has been possible to incorporate both

Fig. 5. Molecular models of (a) lipopolysaccharide from *S.* typhimurium G30 and, (b) lipopolysaccharide and phosphatidylethanolamine. (From ROMEO, GIRARD and ROTHFIELD [8])

the glucosyl and galactosyl transferases (Enz I and Enz II in Fig. 1) into a common monolayer and to demonstrate the sequential transfer of glucose and galactose into the film when the appropriate nucleotide sugars are introduced into the subsolution.

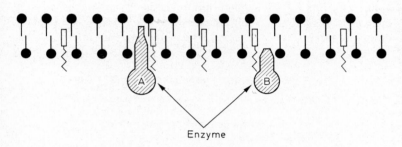

Fig. 6. Diagram of proposed organization of lipopolysaccharide, phosphatidylethanolamine and galactosyl transferase enzyme in the inner membrane of the cell envelope (from ROTHFIELD and ROMEO [15]). A, the enzyme extends into the nonpolar (hydrocarbon) portion of the membrane; B, the enzyme does not extend into the nonpolar region

Membrane Assembly

Although the *in vitro* reassembly system described above certainly does not accurately represent the assembly process *in vivo* since lipopolysaccharide is incorporated into the membrane at an earlier stage of biosynthesis within the cell, the principles emerging from these studies provide a framework for considering events occurring in the *in vivo* assembly process. The following conclusions emerge. (1) The *in vitro* reassembly process is a self-assembly process. The primary structures of the components contain all the information necessary to permit their proper reassociation and no separate catalytic "assembly factor" need be invoked. (2) Reassembly takes place in a stepwise fashion. Therefore a concerted mechanism requiring that all components assemble simultaneously is not required. (3) A sequential order of reassembly was required. The obligatory first step was formation of the mixed monolayer containing lipopolysaccharide and phosphatidylethanolamine which could then readily be penetrated by the transferase enzymes. If the order was reversed, and galactosyl transferase was introduced beneath a monolayer of phosphatidylethanolamine in the absence of lipopolysaccharide, the protein also readily penetrated the film. However, in this case the subsequent introduction of lipopolysaccharide beneath the surface did not result in penetration of the film by the lipopolysaccharide. Therefore, the presence of protein in the lipid monolayer appeared to block access of lipopolysaccharide molecules and prevented formation of the complete system. It has not yet been established whether when the several enzymes of the multienzyme glycosyl transferase system are incorporated, interactions between the proteins introduce new complications into the assembly process. (5) Divalent cations participate in the assembly process at the level of the incorporation of the protein into the mixed monolayer. In the absence of divalent cations, the galactosyl transferase was unable to penetrate the lipopolysaccharide-phosphatidylethanolamine film. Magnesium was the most effective cation but its presence in the subsolution was no longer necessary following introduction of enzyme into the film. Thus, the transfer of galactose from the subphase into the monolayer proceeded in a normal manner when the enzyme-lipopolysaccharide-phosphatidylethanolamine film was moved to the surface of a magnesium-free subsolution prior to introduction of UDP-[^{3}H]galactose. We do not know how much Mg^{++} remained in the film. It has not yet been established whether magnesium acts as a ligand, binding enzyme directly to a component of the film, or whether the cation induces a conformational change in enzyme or film, thereby facilitating their interaction.

Although many questions remain to be answered, it is possible to begin to outline the events that take place during the biogenesis of bacterial and other membranes. These include: (1) *biosynthesis* of the component molecules; (2) *assembly* of component molecules into a common structure; (3) *modification* of the primary structure of membrane components; (4) *translocation* and *migration* of components following their incorporation into the membrane.

(1) Biosynthesis of the lipid components of bacterial membranes appears to take place within the membrane structures themselves. As discussed above, all of the biosynthetic reactions leading to biosynthesis of lipopolysaccharide take place in the inner membrane of the cell envelope, as has been demonstrated by the studies of OSBORN and her co-workers [2]. The presence of the enzymes of

phospholipid synthesis as membrane-bound membranes in both bacteria and higher organisms also suggest that phospholipids are synthesized *in situ*. The site of synthesis of membrane proteins has not been firmly established. It is not known whether membrane proteins are synthesized on cytoplasmic ribosomes and then transported to the membrane or whether some membrane proteins may possibly be synthesized on membrane-bound ribosomes.

(2) The details of the assembly process are still largely unknown due to the difficulty in obtaining purified membrane components suitable for *in vitro* reassembly studies such as those described above. With recent advances which have led to purification of several other membrane enzymes and transport factors, it is likely that a considerable increase in the available information will occur within the next few years.

(3) Modification of the covalent structure of membrane phospholipids occurs at a significant rate in bacteria, with different rates of turnover of the acyl groups and polar portions of the molecules [10]. The deacylation-reacylation cycle of phospholipids in animal cells is also an example of this process [11]. It is possible that similar modifications of membrane glycolipids and glycoproteins may also occur.

(4) Finally, it seems likely that at least some membrane components are capable of movement from their original site of incorporation following the initial assembly process. The three possible types of movement are indicated in Fig. 7.

It is suggested that the term *translocation* be used to described the movement of membrane components from one membrane to another. This clearly occurs in the case of bacterial lipopolysaccharides. OSBORN and her co-workers [2] have shown that the biosynthetic reactions involved in growth of the polysaccharide chain take place in the inner membrane of the cell envelope, followed by the rapid transfer of the newly synthesized molecule to the outer membrane where the bulk of the cellular lipopolysaccharide is located. If this involves passage through the "periplasmic space" which separates the two membranes, a mechanism will be required to mask the hydrophobic hydrocarbon chains of the Lipid A portion of the molecule (see Figs. 1, 5). Translocation of phospholipid molecules also occurs between microsomal and mitochondrial membranes of animal cells, as has been shown by WIRTZ and ZILVERSMIT [13].

A second type of movement involves the migration of molecules within the plane of the membrane. We suggest that the term *cis-migration* be used for this type of movement. In the case of the series of glycosyl transferases (see Fig. 1) which catalyze the sequen-

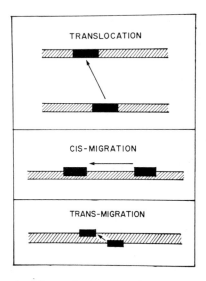

Fig. 7. Translocation and migration of membrane components

tial addition of sugar residues to lipopolysaccharide, extensive studies have failed to reveal release of the enzymes following completion of the transferase reactions, and it appears that the proteins remain within the membrane structure even after catalyzing their respective reactions. Since the lipopolysaccharides which act as substrates for the series of enzymes are also membrane components, it follows that one or both components must migrate within the membrane structure to permit a single molecule of lipopolysaccharide to act as substrate for the series of membrane-bound enzymes. It appears that the most likely candidate for such lateral migration is the lipopolysaccharide since the acyl chains of the Lipid A component are viewed as dissolved in the hydrocarbon interior of the membrane (see Figs. 5 and 6). This should facilitate the lateral diffusion of lipopolysaccharide along the chain of transferase enzymes. The possibility also exists that the inner and outer membranes of the bacterial cell envelope may be continuous in several regions. If this proves to be the case, the translocation of lipopolysaccharide from inner to outer membrane could take place by cis-migration without the need for a mechanism to facilitate transfer of the lipopolysaccharide across the periplasmic space (see above). Other examples of cis-migration may include the rapid redistribution of human and mouse surface antigens following formation of human-mouse hybrid cells by the cell fusion technique [12]. The process of lateral diffusion of individual phospholipid molecules in phospholipid bilayers is also well established.

A third type of movement involves migration of a molecule across the plane of the membrane. It is suggested that this be called *trans-migration*. An example of this process is the transport of molecules across the plasma membrane. In the case of phospholipid bilayers, KORNBERG and MCCONNELL have studied the movement of individual phospholipid molecules from inside to outside surface of the bilayer and have named the process phospholipid flip-flop [16].

The likelihood that membrane molecules can redistribute themselves within the membrane structure following the assembly process would be expected to have major implications for membrane assembly, structure and function.

Acknowledgment

Investigations originating in our laboratory were supported by grants from the American Heart Association and the Health Research Council of the City of New York and by Public Health Service Grants AM-13407 and AM-12711 from the National Institute of Arthritis and Metabolic Diseases.

References

1. HEPPEL, L.: In: The structure and function of biological membranes, Ch. 5. (ROTHFIELD, L., Ed.). New York: Academic Press 1971.
2. OSBORN, M. J., GANDER, J., PARISI, E.: Pers. communication 1971.
3. WESTPHAL, O.: Ann. Inst. Pasteur 98, 789 (1960).
4. OSBORN, M. J.: In: The structure and function of biological membranes, Ch. 7. (ROTHFIELD, L., Ed.). New York: Academic Press 1971.

5. ENDO, A., ROTHFIELD, L.: Biochemistry 8, 3500 (1969).
6. MÜLLER, E., HINCKLEY, A., ROTHFIELD, L.: J. biol. Chem. (in press).
7. WEISER, M., ROTHFIELD, L.: J. biol. Chem. 243, 1320 (1968).
8. ROMEO, D., GIRARD, A., ROTHFIELD, L.: J. molec. Biol. 53, 475 (1970).
9. — HINCKLEY, A., ROTHFIELD, L.: J. molec. Biol. 53, 491 (1970).
10. WHITE, D., TUCKER, A.: J. Lipid Res. 10, 220 (1969).
11. HILL, E. E., LANDS, W. E. M.: Biochim. biophys. Acta (Amst.) 152, 645 (1968).
12. FRYE, L. D., EDIDIN, M.: J. Cell Sci. 7, 391 (1970).
13. WIRTZ, K. W. A., ZILVERSMIT, D. B.: J. biol. Chem. 243, 3596 (1968).
14. BAYER, M. E.: J. gen. Microbiol. 53, 395 (1968).
15. ROTHFIELD, L., ROMEO, D.: Bact. Rev. 35, 14 (1971).
16. KORNBERG, R. D., MCCONNELL, H. M.: Biochemistry 10, 1111 (1971).

Cooperativity in Biomembranes

Donald F. Hölzl Wallach

Tufts New England Medical Center, Boston, MA 02111/USA

With 13 Figures

It is a special honor to give this lecture here, not only because these colloquia have established the highest standards, but also because, as an organizer of the Meeting, I have the pleasure to "stand in" for a most gifted group of thinkers, concerned with one of the most exciting areas of biomedicine. The whole domain of membrane biology is in ferment, as attested by a huge literature, many good, reviews and numerous international conferences. But the phenomenon of *cooperativity* appears common to nearly all membranes, and can be discussed without excessive concern about the molecular architecture of a given membrane. Moreover, the topic links the functionality of membranes to the properties characterizing *regulatory enzymes*.

Cooperative effects in membranes are obvious in many biological events and the matter has been treated brilliantly, from a theoretical point of view, by several workers [1—7], who could, unfortunately, not contribute here. For me, too, it is not a new topic, but I have dealt less with the general problem than with its rather striking manifestations in neoplasia [8—10]. However, here I will address the subject tutorially and emphasize its relevance to the dynamic behavior of cellular membranes, in particular to "all-or-none responses", "triggering" and "gating", as well as to functional pleiomorphism.

First, one must dwell on the physical state of membranes and their symmetry properties: it is a crucial, but often neglected fact that, in their native state, membranes are not the solution systems familar to the biochemist, but essentially solid state assemblies whose components are (a) in high local concentration; (b) in a relatively ordered state, and (c) *locally constrained* (Figs. 1 and 2). The molecular motions accessible to membrane constituents thus differ markedly from the gas or solution state, both in range and frequency. Under all circumstances, the solid state clearly represents greater order and mutual intermolecular constraint than the solution state, and this has major functional and practical implications, e.g.:

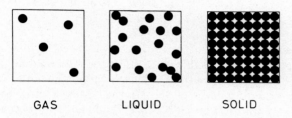

Fig. 1. Molecular "packing" in gas, liquid and solid states

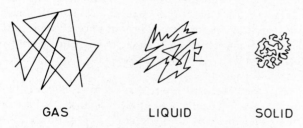

Fig. 2. Molecular motion in gas, liquid and solid states

(a) If a membrane protein is synthesized in *water-soluble form* by familiar mechanisms, how does it attain its membrane location, and once there, how does the molecule in its *membrane location* resemble its possible state in solution?

(b) If we *extract* macromolecules from a membrane, e.g. by ionic manipulations, detergents or lipases, how does their extracted state resemble that *in situ*?

These are important unsolved technical questions in membrane biology, but they will not be dealt with further here.

The degree of order in the solid state can vary markedly and there is increasing evidence that substantial areas of membrane surfaces constitute *highly ordered lattices*. These are observed more and more often by micromorphologic methods. There are many

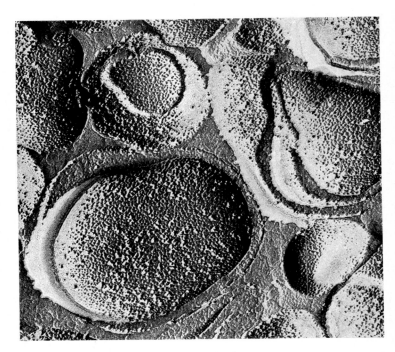

Fig. 3. Lattice-structuring in erythrocyte membranes revealed by "freeze-etch" electron microscopy. Note the ordered appearance at the fracture-face, underlying the scattered "membrane-associated particles". Mag. 120,000. Courtesy of V. SPETH

examples [11] and among the best are the so-called "tight junctions", seen to occur between many cell membranes, and likely sites for intercellular molecular channels, allowing passage of molecules of up to 10,000 daltons.

Careful reconstruction of such membrane regions [12] indicates hexagonal lattices both within the participating membranes and the local intercellular material, with possible hydrophilic channels running from cell to cell. The latter might be analogous to (but larger than) the pores penetrating the hemoglobin tetramer [13]. Such lattices occur also in other membrane regions [11] and a striking example from human erythrocytes is shown in Fig. 3. Such lattices are unlikely to be protein only, consisting rather of proteins associated with their characteristic lipids, perhaps as a

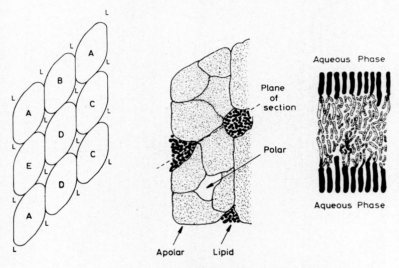

Fig. 4. Mosaic membrane model. See text and [14] for further details

mosaic, such as is conceptualized in Fig. 4. This illustration, deriving from a previous model developed in our group [14], focuses also on the modes of lipid-protein interaction in membranes, emphasizing the hydrophobic interrelations between these membrane components and the structural perturbations likely to be introduced in purely lipid domains by the protein, but at a considerable distance from it.

The diagram shows an *asymmetric, mosaic lattice* involving diverse lipo-protein subunits, the whole exhibiting the transverse and tangential polarities suggested by CHANGEUX et al. [1—3] and

theoretically shown to be essential to membrane coupling mechanisms [15]. That *transverse* polarity exists is now established by:
(a) *binding* studies, e.g. those showing that the binding sites for tetrodotoxin occur only on the outer neuronal surface [16];
(b) our own work [17, 18] on the accessibility of the membrane proteins of erythrocyte membranes to protease action and from the two sides of the membrane, conceptualized in Fig. 5.

Tangential polarity has been less obvious in animal membranes, but there is indirect, functional evidence for it. It is also strikingly

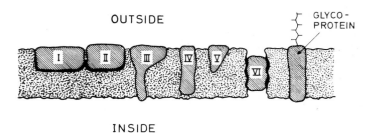

Fig. 5. Transverse polarity in isolated erythrocyte membranes, as revealed by accessibility of major membrane peptides (I—VI) to protease action from the two sides of the membrane (from ref. [18])

apparent in ciliate protozoa [19, 20] and it has been illustrated for lymphocytes by HÄMMERLING in this symposium.

Fig. 5 symbolically localizes the *transverse* polarity of the major peptides in human erythrocyte membranes. It derives from the accessibility of the various membrane peptides to trypsin, chymotrypsin, papain, etc., applied to the inside and outside surfaces of membrane vesicles impermeable to the proteases [18]. "Inside out" and normally-orientated membrane vesicles derived from erythrocyte "ghosts" [21] were used to achieve this spatial discrimination. No implications as to the distribution of these peptides within the plane of the membrane are intended. Also, one must recall that the diagram applies only for purified, hemoglobin-free erythrocyte membranes, which may not truly represent the state of the membranes in intact erythrocytes. (For example, intact erythrocytes are very resistant to external application of proteases, whereas this is clearly not so for isolated membranes [18]). In any event, in purified erythrocyte membranes all of the detectable peptides, including the glycopeptides, appear to exhibit transverse polarity. Moreover, some at least appear to penetrate the membrane. Possibly all of the peptides are represented on both membrane surfaces, but components I, II, IV and V of Fig. 5 are resistant to proteolysis from "inside".

In Fig. 6, the major peptides of erythrocyte membranes of several species are shown separated according to their *molecular weight* by electrophoretic molecular sieving in polyacrylamide containing SDS and dithiothreitol [22]. The patterns are rather similar and about eight bands are consistently seen with protein stains and, in human membranes, components I—VI account for about 65% of the staining with several protein reagents [17]. There are three glycoproteins, accounting for about 5% of the membrane protein.

The broad range of molecular weights (more than 150,000 to about 30,000) and the apparent presence of high order in erythrocyte membranes (Fig. 3) creates a dilemma:

(a) either the various peptides (here separated artificially in detergents) somehow assemble as exact structural equivalents in the intact membrane, or

(b) the absolute identity and symmetry expected from oligomeric proteins is replaced in membranes by certain structural homologies yielding an *equivalence of association*, as suggested by CHANGEUX and associates [1—4]. In this way the requirements for an ordered structure are maintained, although the subunits are not totally identical.

The assembly of subunits of equivalent association into effectively infinite 2-dimensional topology of a membrane carries major functional implications, because (1) membranes are biologically "excitable" in that they respond to *regulatory ligands* (non-catalytic and not bound covalently) by a change in *state* and in the *activity* of their functional components and (2) most membranes exhibit an *apparent cooperativity* in their biological activity, depending upon *poising* levels of regulatory agents ("ligands"); this provides some membranes with an "amplifying" ability which can even yield an

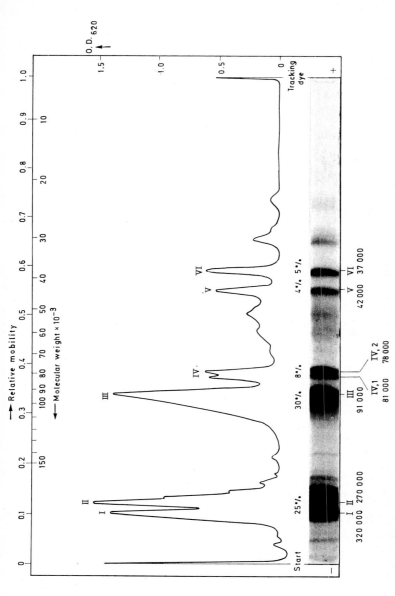

Fig. 6. Major peptides of erythrocyte membranes of several species, separated according to their molecular weight by electrophoretic molecular sieving in polyacrylamide. (H. KNÜFERMANN and D. F. H. WALLACH, unpublished)

"all-or-none" response. Some possible regulatory agents are listed in Table 1 (cf. also [23]).

Before examining cooperative effects in membrane, it is useful to look at the cooperative behavior of a soluble oligomeric protein, such as hemoglobin (Figs. 7 and 8). Myoglobin (Mb), which can be

Table 1

Membrane ligands

1. *Molecular weight less than 50 Daltons*
   H^+, K^+, Na^+, Ca^{++}, etc.

2. *Molecular weight 50 to 1000 Daltons*
   a) Phospholipids, Steroids
   b) Acetylcholine, Norepinephrine, etc.
   c) Antibiotics
   d) Sugars, Aminoacids, etc.

3. *Macromolecules*
   a) Enzymes
   b) Protein hormones
   c) Antibodies
   d) Antigens
   e) Lectins
   f) Complement components
   g) Nucleic acids
   h) Colicins
   i) Hemoglobulin (in erythrocytes)
   j) Other soluble "Cytoplasmic" Proteins

4. *Supramolecular complexes*
   a) Viruses
   b) Polysomes
   c) Other cells

viewed heuristically as a hemoglobin protomer, behaves in a simple fashion, its single heme reacting with a single ligand molecule in a straightforward fashion. Hence the binding curve of Mb for oxygen (Fig. 7) describes a simple hyperbola. However, in hemoglobin (Hb), the four protomers interact so as to let the molecule release oxygen efficiently at low oxygen concentration (tissues), and to scavenge it effectively at the high oxygen levels existing in the lungs. The

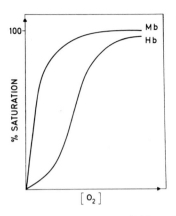

Fig. 7. Oxygen binding curves of myoglobin and hemoglobin

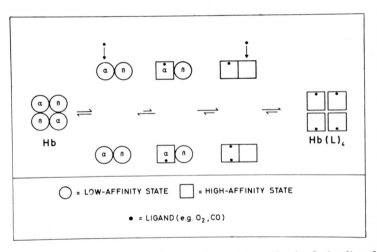

Fig. 8. Cooperative interactions between hemoglobin subunits during ligand binding. See text

oxygen binding curve of Hb thus exhibits a "sigmoid shape" (Fig. 7) arising from the cooperative interactions between α and β protomers schematized in Fig. 8. In the hemoglobin case, subunit association actually "restrains" the intrinsic (microscopic) liganding activity of the prosthetic groups of individual protomers. Together with ligand binding there occur changes in tertiary and quaternary structures, especially in subunit relationships; these have long been noted biochemically, and have been mapped by X-ray crystallography [24].

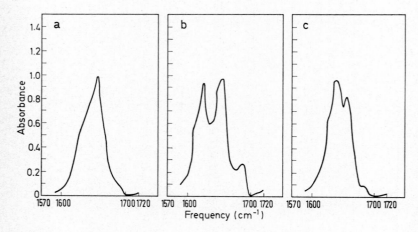

Fig. 9. Spectral changes in the Amide I region of erythrocyte ghosts associated with ATP hydrolysis. a) Control; b) + ATP-Mg; c) + ATP-Mg in presence of Na^+ and K^+. Spectra are of membranes in D_2O suspension and indicate a shift of peptide conformation to antiparallel β-structure with ATP-hydrolysis [26]

In membranes also, one gets changes in *molecular architecture* and function as a result of binding suitable ligands. These emerge perhaps most strikingly from our own spectroscopic studies of mitochondrial and erythrocyte membranes during biological activity [25, 26]. Our infrared (IR) spectra show that the proteins of *mitochondrial* membranes usually contain an appreciable proportion of their peptide in the antiparallel β-conformation. This proportion increases markedly with rate of electron transport and declines when this is coupled to oxidative phosphorylation [25]. "Energization" of *erythrocyte* "ghosts" by ATP-Mg (Fig. 9) produces a transition to the antiparallel beta structure also in the peptides of these membranes, to an extent depending on the rate of ATP hydrolysis. "Resting" erythrocytes exhibit no β-structure, but this conformation becomes dominant upon activation of the Na^+-K^+-stimulated ATPase [26]. The amplitude of the peptide rearrangements reported by IR spectroscopy remains to be determined. It is also not known whether they reflect local changes of pH or electric field, or whether they represent cooperative changes. Nevertheless, since these and other data show that changes of molecular structure occur within membranes under appropriate physical conditions and stimuli, one must delve into possible, unusual consequences of such changes in a *lattice system*. The *first* of these is that the *binding of a structure-determining ligand* to one of the many *subunits of equivalent association may perturb other subunits with other functions and thus produce diverse and multiple functional manifestations* — i.e. *functional pleiotropism* [8—10]. The *second* consequence is equally crucial and still largely hypothetical — i.e. membranes are expected to be *highly cooperative* systems. Here I will only present a tutorial overview of some of the theoretical approaches developed by CHANGEUX and associates [1—4], leading to a likely experimental example, coming from METCALFE's [27] work: the last also introduces some of the future experimental approaches to this problem.

It is important to recall that in ordinary, dilute solutions, a soluble enzyme may change shape with only minor restraints from the solvent molecules which surround it. Such reactions are likely to be kinetically and otherwise modified in *membranes* because of association of the proteins with other membranes and/or structured lipid aggregates. As in simpler cases, e.g. hemoglobin, the forces which constrain a single unit in an array can yield cooperativity of the system as a whole. Because macromolecular interactions in membranes are extensive, cooperative effects are potentially large and may be *propagated* over wide distances — like *lattice defects* in a crystal.

In any event, one can envisage that each subunit can exist in at least two structural states, R and S, which relate through a reversible equilibrium:

$$R \rightleftharpoons S$$

It is also reasonable to assume that the ligand affinities for R and S are different, i.e. that
$$RL \leftrightharpoons R + L; K_R$$
and
$$SL \leftrightharpoons S + L; K_S$$
where K_R and K_S are "overall" dissociation constants and $K_R < K_S$ in a positively cooperating system.

Two possibilities need to be considered:

(a) that interactions proceed principally through *indirect coupling*, e.g.

Receptor $\xrightarrow{1}$ Transducer $\xrightarrow{2}$ Amplifier $\xrightarrow{3}$ Membrane effector.

Such a multistep process, where the enzyme adenyl cyclase acts as amplifier, the structural reorganization of the membrane resulting through the action of a *diffusible* regulator, cyclic AMP, has been proposed by RODBELL [28] and is discussed in [3] and by SCHMITT in this symposium.

(b) that *structural coupling* is important.

The cooperative, mutual interactions envisaged in this case mean that the R — S transitions of each subunit depend upon the structural state of its neighbors; this mechanism is very useful in the explanation of cooperativity and functional pleiotropism in membranes. In this model, the functional subunits may be (1) distributed irregularly over the membrane surface as *small, localized, oligomeric clusters* or (2) they may be part of a *lattice* with equivalent associations between neighbors. As pointed out by CHANGEUX and associates [1—4], only the second case can explain *both* the graded and all-or-none responses observed in membranes, as well as their functional pleiomorphism [8—10].

Exact descriptions of the behavior of such a two-dimensional lattice model are not yet available, but CHANGEUX and associates [1—4] have used the "molecular field" of the BRAGG-WILLIAMS approximation to yield a useful heuristic treatment of the situation (Fig. 10).

Consider first the transformation of one lattice subunit from the S to the R state, when all other subunits are in the S state. The free energy of the transition ε and the apparent isomerization constant $I = (S/R)$ for the transition can be related as follows:
$$I = \exp(N/RT \cdot \varepsilon).$$

The situation will be distinctly different in an S ⇌ R transition when the lattice already contains some subunits in the R state. Then, the promotion energy $\Delta F = \varepsilon'$ for a given protomer to

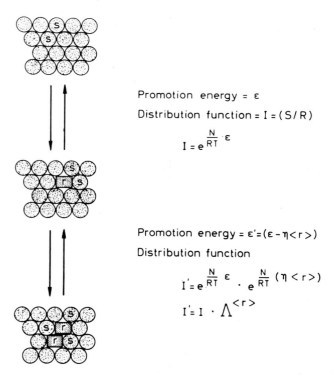

Fig. 10. Cooperativity in a membrane lattice (see text)

undergo the S — R transition will be linearly related to the fraction $\langle r \rangle$ of subunits which are already in the R state. Then

$$\varepsilon' = \varepsilon - \eta \langle r \rangle$$

and the new *apparent isomerization constant* I' will be

$$I' = \exp(N/RT)\, \varepsilon - \eta \langle r \rangle$$

or

$$I' = I \cdot \Lambda^{\langle r \rangle}.$$

Here R = the gas constant, N = Avogadro's number, T = absolute temperature, η describes the interaction energy between nearest neighbors and A is a cooperativity parameter. When $\varLambda = 1$, there is no cooperativity and decreasing values of $\varLambda$ signify increasing positive interaction.

Numerical evaluation of the above equations allows one to plot (1) a *state function* $\langle r \rangle$ and (2) a *binding function* $\langle y \rangle$ vs $[L]/K_r$, for varying values of $\varLambda$, where $\langle r \rangle$ is the proportion of subunits in the R-state, $\langle y \rangle$ the proportion of subunits combined with the ligand L, and [L] the ligand concentration. Such curves are given for two values of the cooperative factor $\varLambda$ in Figs. 11 and 12. In Fig. 11 (low cooperativity) $\langle y \rangle$ vs $[L]/K_R$ resembles the sigmoid ligand-binding curve of hemoglobin. However, the $\langle r \rangle$ and $\langle y \rangle$ functions trace dissimilar courses, and $\langle r \rangle$ is not 0 when [L] = 0. This is an implicit consequence of the contention that the R ⇌ S equilibrium can exist *without* ligand binding, and distinguishes the model of CHANGEUX and associates from "induced fit" models [29].

An important consequence of the independent course of $\langle r \rangle$ and $\langle y \rangle$ is that, at certain concentrations of L, small changes of state can effect the binding or unloading of large amounts of ligand or, conversely, small changes in ligand binding can cause large changes of state. This "amplifying" property of the model is diagrammed in Fig. 11 (curve a), deriving from the mathematical formulations of CHANGEUX and THIERY [2], who define the amplifying factor a as

$$a(\alpha) = \frac{1}{\langle r \rangle_{max} - \langle r \rangle_{min}} \frac{\partial \langle r \rangle}{\partial r}$$

where $\alpha = L/k_R$ and k_R is the microscopic dissociation constant of the R-state for bound ligand L. The amplification factor clearly goes through a sharp maximum at a defined value of L, and slightly lower ligand concentrations can be considered as "poising" levels for the system.

It is also profitable to consider a case of high cooperativity, i.e. where $\varLambda$ is small, and when the amplification becomes infinite at certain values of $[L]/K_R$ and the system becomes metastable (Fig. 12). Then a very small change of [L] and the binding of only a few molecules of ligand can perturb the whole membrane lattice, and yield at the same time multiple alterations of membrane functions "linked" through the equivalent associations of the membrane subunits.

All of this clearly depends on the parameters [L], K_R and Λ. [L], which means effectively that the steady state of the regulatory ligand is presumably subject to controls *both* internal and external

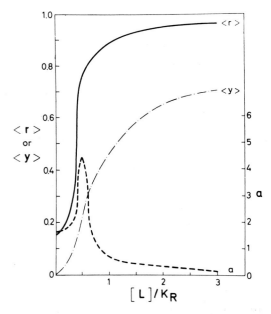

Fig. 11. State function $\langle r \rangle$, binding function $\langle y \rangle$ and "amplification" as a function of $[L]/K_R$ in a cooperative membrane lattice. See text

i.e. genetic; and enzymatic (where L is a metabolic intermediate), or by the rate of production and degradation of a regulatory hormone, etc. Moreover, such external mechanisms may normally maintain [L] at a "poising" level so that minor changes in L, brought about by some biologic stimulus, can effect very large changes in state, as suggested by the "all-or-none", "triggering" and "gating" effects commonly observed in membranes. Experimental verification of this model will not be easy to achieve, but I will cite an experiment of METCALFE [27] which may show such direct cooperative effects in a membrane lattice system; they have the weakness, however, of not being reversible in *practice*.

It is well established that many agents which at high concentrations disrupt erythrocyte membranes, at low levels stabilize

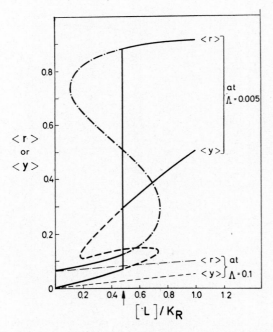

Fig. 12. Discontinuities in ⟨r⟩ and ⟨y⟩ possible in highly cooperative lattices at critical values of $[L]/K_R$. See text

them against osmotic and mechanical stress. The switch from stabilization to disruption typically occurs over a narrow concentration range of the membrane-active agents, and this "critical" concentration can vary widely from one substance to another.

This behavior is exemplified by benzyl alcohol, BeOH, as shown in the upper panel of Fig. 13, although this substance is lytic only at rather high bulk levels. However, BeOH has the advantage that its benzyl protons exhibit a characteristic magnetic resonance (PMR), so that it can *both* perturb the membrane and record the nature of the perturbation (Fig. 13, lower panel). During stabilization, the bandwidth of the benzyl proton spectrum de-

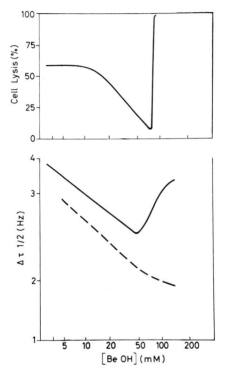

Fig. 13. Discontinuities of membrane state, revealed by hemolysis (upper panel), and ligand binding, revealed by PMR bandwidth (lower panel), at increasing levels of benzyl alcohol. See text. From [27]

creases progressively, suggesting partition of BeOH from the aqueous phase into lipophilic membrane domains. This behavior is mimicked by membrane lipids in water dispersion.

Importantly, over a narrow range of the ligand concentration BeOH one observes: (1) a discontinuous change of *membrane state* — lysis, and (2) a discontinuity in the benzyl PMR bandwidth, signifying an abrupt increase in the *binding* of the ligand, BeOH.

These phenomena are consistent with the hypothesis of CHANGEUX and associates (although they do not prove it) and point to a potential method of approach to this important field.

The biomedical behavior of lattice-ordered biomembranes is indubitably of major interest and could be altered in a cooperative fashion through at least five general mechanisms:

1) by introduction of a "new" membrane subunit, via mutation, as a viral gene product, or by modification of existing subunits by ionizing radiation, etc.;

2) by a change in the steady-state concentration of a "native", structure-determining ligand;

3) by the appearance of an isomer of a structure-determining ligand through mutation or viral infection; certain neuroactive drugs, including addictive and hallucinogenic agents, might function in this manner;

4) by exposure to a foreign ligand with high membrane affinity, e. g. antigens, lipophilic drugs, colicins;

5) by alteration of existing membrane components through external lytic enzymes, or activation of intrinsic ones. Membrane penetration by large viruses, sperm and parasites might be considered from this view.

Acknowledgement

The author acknowledges the support of the Max Planck Society, the John S. Guggenheim Memorial Foundation and the Andres Soriano Cancer Research Foundation.

References

1. CHANGEUX, J.-P., THIERY, J., TUNG, Y., KITTEL, C.: Proc. nat. Acad. Sci. (Wash.) **57**, 335 (1967).
2. — — In: Regulatory functions of biological membranes (JÄRNEFELT, J., Ed.). B. B. A. Library 11, p. 115. Amsterdam: Elsevier 1968.
3. — In: Nobel Symp. No. 11. Symmetry and function of biological systems at the molecular level, p. 236 (ENGELSTRÖM, A., STRANDBERG, B., Eds.). New York: John Wiley 1969.
4. BLUMENTHAL, R., CHANGEUX, J.-P., LEFEVER, R.: J. Memb. Biol. **2**, 351 (1970).
5. WATANAKE, A., TASAKI, I., LERMAN, L.: Proc. nat. Acad. Sci. (Wash.) **58**, 2246 (1967).
6. HILL, T. L.: Statistical mechanics. New York: McGraw-Hill 1956.
7. — Proc. nat. Acad. Sci. (Wash.) **58**, 111 (1967).

8. WALLACH, D. F. H.: Proc. nat. Acad. Sci. (Wash.) **61**, 868 (1968).
9. — New Engl. J. Med. **280**, 761 (1969).
10. — Curr. Top. Microbiol. Immunol. **47**, 152 (1969).
11. MOOR, H.: Int. Rev. expte. Path. **5,** 179 (1966).
12. MCNUTT, N. S., WEINSTEIN, R. S.: J. Cell Biol. **47**, 666 (1970).
13. PERUTZ, M. F.: J. molec. Biol. **13**, 646 (1965).
14. WALLACH, D. F. H., GORDON, A. S.: Fed. Proc. **27**, 1263 (1968).
15. KATCHALSKY, A., CURRAN, P. F.: In: Membrane biophysics, p. 88. Cambridge, Mass.: Harvard University Press 1965.
16. NAHARSHI, T., MOORE, J. W.: J. gen. Physiol. **51**, 935 (1968).
17. FAIRBANKS, G. F., STECK, T. L., WALLACH, D. F. H.: Biochemistry (1971) (in press).
18. STECK, T. L., FAIRBANKS, G. F., WALLACH, D. F. H.: Biochemistry (1971) (in press).
19. CHATTON, E., LWOFF, A.: C. R. Soc. Biol. (Paris) **104**, 834 (1930).
20. BUSSON, J., SONNENBORN, T. M.: Proc. nat. Acad. Sci. (Wash.) **53**, 275 (1965).
21. STECK, T. L., WEINSTEIN, R. S., STRAUS, J. H., WALLACH, D. F. H.: Science **168**, 225 (1970).
22. KNÜFERMANN, H., WALLACH, D. F. H.: To be published.
23. WEISS, P.: Proc. nat. Acad. Sci. (Wash.)
24. MUIRHEAD, H., COX, J. M., MAZARELLA, PERUTZ, M.: J. molec. Biol. **28**, 117 (1967).
25. GRAHAM, J. M., WALLACH, D. F. H.: Biochim. biophys. Acta (Amst.) **193**, 225 (1969).
26. — — Biochim. biophys. Acta (Amst.) 1971 (in press).
27. METCALFE, J. C.: This symposium.
28. RODBELL, M.: Quoted by F. D. SCHMITT in this symposium.
29. KOSHLAND, D. E.: In: The Enzymes, 2nd Ed., Vol. I, p. 305 (BOYER, P. P., CARDY, H., MYRBÄCK, K., Eds.). New York: Academic Press 1959.

Magnetic Resonance Studies of Membranes and Lipids

J. C. METCALFE

Medical Research Council, Molecular Pharmacology Unit, Medical School, Hills Road, Cambridge CB2 2QD./GB

With 21 Figures

To relate the structure of a membrane to its functional properties at the molecular level requires techniques to define the detailed organization of the membrane components, and to identify the components of the structure involved in a specific function in the intact membrane. Both problems are complicated by the multiplicity of the membrane components. The use of X-ray diffraction techniques to define the spatial relationships of the membrane components has been complicated by the dynamic motion of the membrane lipids and the absence of a high degree of order in the membrane proteins, so that the diffraction patterns contain only limited structural information. The techniques of nuclear magnetic resonance (NMR) and electron spin resonance (ESR) are more appropriate for a dynamic structure, in which both the detailed molecular motion of the components and their average distance of interaction can, in principle, be defined. The underlying assumption in this approach is that the assembly of the components into the membrane structure imposes mutual steric restrictions on the components which will be expressed as a characteristic pattern of molecular motion. Since there is also good evidence that functional proteins in a number of membranes have precise lipid requirements, it would be expected that functional membrane proteins would impose characteristic dynamic patterns on the lipids with which they interact.

We outline the measurement of the dynamic properties of lipid molecules in vesicles of lipid bilayers by the magnetic resonance techniques and examine the effect on these dynamic properties of extraneous molecules which alter the permeability of the vesicles to solutes. The key question is whether the dynamic perturbations induced by the extraneous molecules determine the simultaneous permeability changes; that is, whether we are looking at a relevant structural perturbation.

The simple lipid vesicles provide an essential baseline for the extension of those measurements to biological membranes, for which techniques of specific isotopic substitution are essential to simplify the spectra to known chemical components in the membrane. Finally, we review earlier probe magnetic resonance experiments which give more limited but clear-cut information about interactions between components in the intact membrane structure.

The Dynamic Properties of Membrane Lipids

The dynamic properties of lipids which can be measured by magnetic resonance techniques (NMR and ESR) are:

(i) The motion of the chains through the thickness of the bilayer.

(ii) The rate at which the lipid molecules diffuse laterally in the surface of the bilayer, i.e. the rate at which nearest neighbors exchange positions. Only order of magnitude limits can be set to this diffusion rate at present, and although it is an important parameter, further discussion is deferred until the technical problems of measurement have been solved.

(iii) The rate at which lipids flip from one side of the bilayer to the other.

Chain Motion

The measurements are illustrated for sonicated vesicles of egg lecithin or dipalmitoyl lecithin of ca. 250 Å diameter. The molecular motion of the chains in bilayers has been measured by both NMR [1, 2] and spin labeling techniques [3]. The proton spectrum of lecithin in $CDCl_3$ is shown in Fig. 1 (a) with the $^+NMe_3$ and alkyl chain ($[CH_2]_n$ and $-CH_3$) resonances identified. The spectrum is relatively sharp because of the rapid molecular motion in this solvent, but in the bilayer structure in D_2O the resonances of the fatty acid chains are broadened by the steric interactions of the chains packed together laterally. Because of the overlap of the $(CH_2)_n$ protons we cannot follow chain motion along the chain in detail even in this simple system by proton NMR. It is more informative to use the ^{13}C NMR spectrum of the lipid [1], in which six carbons can be observed from the fatty acid chains in dipalmitoyl lecithin in addition to the main $(CH_2)_n$ envelope containing 10 carbon nuclei. These carbon nuclei remain well resolved in the bilayer structure in D_2O (Fig. 2), allowing relaxation measurements to be made on each resolved resonance in the structure. The T_1 relaxation times of these carbons in the bilayer are a measure of their rate of molecular motion, and there is a 30-fold increase in T_1 from the carboxyl end of the chain to the terminal methyl (Table 1), implying a large increase in

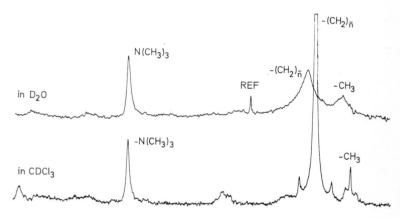

Fig. 1. Proton NMR spectra of lecithin at 30 °C in (a) $CDCl_3$, (b) D_2O

motional freedom along the chain. The $+N(CH_3)_3$ carbons have T_1 values comparable with the main $(CH_2)_n$ envelope so that molecular motion also increases from the glycerol bridge region of the structure out towards the polar interface. To obtain a complete description of the fatty acid chain motion requires specific ^{13}C enrichment of the appropriate carbon nuclei. A full relaxation curve will provide a precise description of chain motion which is likely to be highly sensitive to interaction with other membrane components.

A striking difference in the relaxation times in $CDCl_3$ compared with D_2O is the very short value for the $+NMe_3$ carbon nuclei relative to the $(CH_2)_n$ resonance envelope in $CDCl_3$ (Table 1). This is probably due to the formation of a micellar structure with the polar headgroups restricted in molecular motion by being at the center of the micelle with the alkyl chains exposed to the $CDCl_3$ solvent. This is in contrast to the bilayer structure in D_2O where the headgroups are exposed to the D_2O solvent which allows relatively free motion.

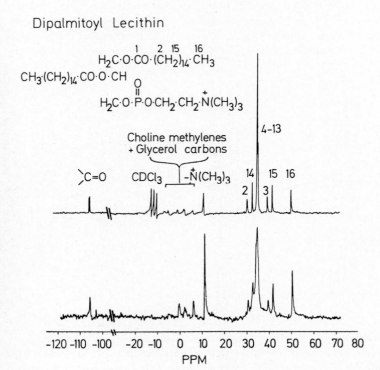

Fig. 2. ^{13}C NMR spectra of dipalmitoyl lecithin in (a) $CDCl_3$ at 28 °C, (b) D_2O at 52 °C

Preliminary data has been obtained from similar experiments using specifically fluorinated stearic acid derivatives, labeled at three positions along the chain replacing single protons [2]. These monofluorostearic acids were incorporated into lecithin bilayers (5 lecithin: 1 fatty acid molecule) and the linewidths measured.

There is substantial broadening of the resonance from the ^{19}F nucleus in the bilayer and we take the broadening as a measure of the relaxation rate ($1/T_2$) and the motional restriction on the chain. The linewidth measurements for the three fluorostearic acids in the bilayer as a function of temperature are shown in Fig. 3a. The linewidths decrease markedly as the ^{19}F nucleus is moved towards the terminal methyl group. From the limited data available, the line-

Table 1

Carbon	T_1 (sec)	
	D_2O (52°)	$CDCl_3$ (28°)
2	0.10	0.19
3	0.22	0.28
4—13	0.55	1.04
14	1.14	2.20
15	1.76	2.65
16	3.34	2.88
$-\overset{+}{N}(CH_3)_3$	0.70	0.19

T_1 relaxation times of carbon nuclei in dipalmitoyl lecithin in D_2O and $CDCl_3$. Palmitic acid carbons are numbered from the carboxyl group [1] to the terminal methyl [16].

width change is approximately linear with substituent positions (Fig. 3b), but the complete form of the curve for the bilayer will only be defined when data is available for the full length of the chain in ^{19}F labeled lecithins. These molecules provide an ideal system for examining the relationship between the two NMR relaxation times T_1 and T_2 along the chain, which is important in the theoretical treatment of the relaxation in relation to chain motion. There is little doubt that at least qualitatively the descriptions of chain motion by both the ^{13}C and ^{19}F NMR experiments are consistent with each other, and with the spin label experiments of HUBBELL and McCONNELL [3].

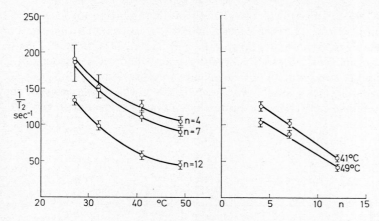

Fig. 3. Relaxation rates ($1/T_2$) of ^{19}F nuclei in monofluorostearic acids incorporated into lecithin bilayers as a function of (a) temperature, (b) ^{19}F substituent position. The carbons are numbered from the carboxyl group ($n = 1$) in all figures

They used lecithin molecules carrying the nitroxide group at different positions along the fatty acid chain, analogous to the ^{19}F labeled analogues above:

The ESR spectra of these nitroxides are also sensitive to molecular motion; the spectrum in $CDCl_3$ is a symmetrical triplet, but when inserted into lecithin vesicles in an aqueous suspension, the spectrum is broadened and asymmetric due to the restriction on motion in the bilayer structure (Fig. 4). From the form of the curves and, in particular, the positions of the extrema, it was shown that the lecithin molecule in the bilayer is tumbling very rapidly about its long axis ($\tau_c \sim 10^{-9}$ sec), whereas the motion about the other two axes is relatively slow, so that the overall motion of the nitroxide on the lecithin chain is severely anisotropic. In a recent refinement of this model McFarland and McConnell [4] have interpreted the spectra to show that the chains are tilted for the first section of the carbon chain from the acyl group at a calculated angle of 60° to the surface of the bilayer, and are also very tightly packed in this tilted region of the structure. In separate experiments it has been

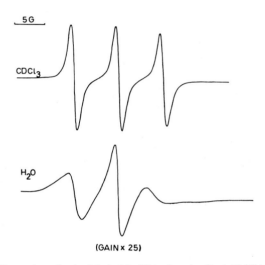

Fig. 4. ESR spectra of spin-labeled lecithin (see text) at 25 °C in (a) $CDCl_3$, (b) D_2O, incorporated into lecithin vesicles

shown that the long axis of the molecule is oriented *on average* perpendicular to the surface of the bilayer [5]. The tilted packing arrangement, first proposed on an experimental basis by Levine and Wilkins [6] from X-ray diffraction data, allows sufficient area per molecule to accommodate the increase in motion along the chain implied in the ESR spectra in Fig. 5. As the nitroxide group is attached further away from the carboxyl group in the bilayer, the spectra become progressively sharper and the extrema which characterise the anisotropy of the motion move inwards and eventually disappear. These effects were quantitated

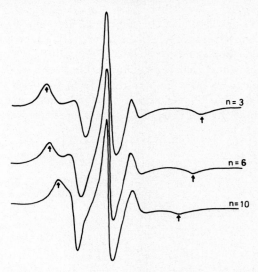

Fig. 5. ESR spectra of spin-labeled lecithins in lecithin-cholesterol (2:1) bilayers, for different nitroxide substituent positions in the fatty acid chain (n = 3, 6, 10). Adapted from HUBBELL and McCONNELL [3]

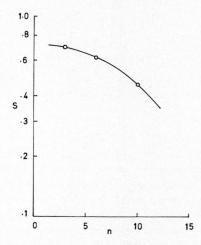

Fig. 6. Order parameter, S, of spin-labeled lecithins in lecithin-cholesterol (2:1) bilayers for different nitroxide substituent positions in the fatty acid chain. Adapted from HUBBELL and McCONNELL [3]

by HUBBELL and MCCONNELL [3] as an order parameter S, which decreases with increasing distance from the carboxyl group (Fig. 6). The simplest interpretation of the order parameter is that it is a measure of the probability of finding all the carbons in the chain up to the nitroxide group in the fully extended (all trans) conformation without a gauche conformation present ($S = 1$).

The bilayer therefore appears to have a relatively fluid interior, with a tightly packed structure at the glycerol bridge, as evidenced by the increased motional freedom from the glycerol group towards the terminal methyl of the fatty acid chain and towards the NMe_3 of the polar head group.

The Effect of Perturbing Agents on Vesicle Permebaility

It is well known from the work of VAN DEENEN et al. [7] and KLINE et al. [8] that the structure of the phospholipid itself has a large effect on the permeability properties of the vesicles. For example, if the polar head group of the phospholipid is choline, and the structure of the fatty acid chains is systematically varied in the lecithin molecules, the permeability of glycerol into the vesicles increases as the chain length is decreased, increases with the introduction of double bonds, and increases if the lengths of the two chains are disproportionated, keeping the total number of carbons constant. On the other hand, the permeability decreases with the addition of cholesterol. Recently, VAN DEENEN [7] has also shown that similar qualitative changes in permeability were obtained in Mycoplasma laidlawii membranes as the phospholipid structure was varied through the fatty acid composition of the growth medium.

This data suggests that the permeability changes with lipid structure are related to how the fatty acid chains pack together in the bilayer, and we would expect this packing to be expressed in the molecular motions of the chains. Here we use two perturbing agents which have large and opposing effects on permeability to illustrate the effects on chain motion; cholesterol, which decreases the permeability of the vesicles, and benzyl alcohol which causes a large increase in permeability. Examples of these permeability changes are shown in Fig. 7 where the penetration of ascorbate into lecithin vesicles has been measured by a spin label reduction technique described later [9]. Benzyl alcohol causes an exponential increase in penetration rate, and this effect is strongly antagonised by cholesterol.

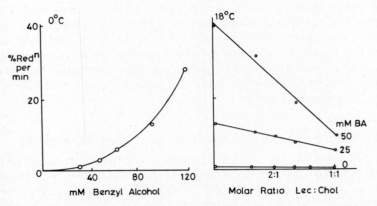

Fig. 7. Permeability of ascorbate ions into lecithin vesicles measured by a spin label reduction technique (see text). (a) Effect of benzyl alcohol at 0 °C, (b) Effect of cholesterol and benzyl alcohol at 18 °C

Table 2

Sample	T_1 (sec)		
	$^+NMe_3$	$(CH_2)_n$	$-CH_3$
Lecithin	0.39 ± 0.01	0.47 ± 0.02	0.54 ± 0.02
2:1 Lecithin:cholesterol	0.35 ± 0.01	0.14 ± 0.03	0.19 ± 0.02

T_1 relaxation times of proton nuclei in egg lecithin in D_2O at 40 °C.

The effect of cholesterol on proton relaxation times (T_1) for lecithin bilayers is shown in Table 2 [10]. The effect of cholesterol (2 lecithin: 1 cholesterol) is to cause a large decrease in T_1 for the terminal methyl and $(CH_2)_n$ resonances by a factor of ca. 3, but a much smaller decrease in T_1 for the $^+NMe_3$ in the headgroup region. These results are consistent with intercalation of cholesterol with the fatty acid chains causing increased packing and immobilization, and a small effect on the $^+\text{NMe}_3$ group transmitted from the chains. So far, there is no evidence for a direct interaction of the cholesterol hydroxyl group with the choline phosphate moiety. In contrast, benzyl alcohol causes a substantial increase in T_1 for the $^+\text{NMe}_3$ group but has only a very small effect in the same direction on the terminal methyl and $(\text{CH}_2)_n$ relaxation, consistent with a fluidising or disordering effect of the alcohol primarily localized in the polar region of the structure.

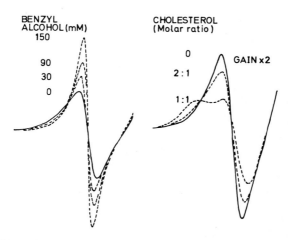

Fig. 8. Effect of (a) benzyl alcohol, (b) cholesterol on the ESR spectrum of a stearic acid spin label (n = 7) in lecithin vesicles at 25 °C

The corresponding spin label experiments generally confirm these NMR results. Fig. 8 shows the spectrum of a stearic acid spin labeled in the chain (n = 7), which is progressively sharpened by increasing benzyl alcohol concentrations, indicating a fluidising effect in the bilayer, while cholesterol has a pronounced effect in the opposite direction [9].

Using a lecithin spin label first synthesized by KORNBERG [11] with a quaternary methyl of the headgroup replaced by a tempoyl group, it was found that benzyl alcohol sharpens the spectrum (Fig. 9) as expected from the NMR data, but cholesterol also has the same effect. In the presence of cholesterol, benzyl alcohol has a negligible effect on the spectrum. We think that this discrepancy with the NMR data arises because the tempoyl group is much larger than the methyl group it replaces, and that its steric interactions with the neighbouring lecithin $^+NM_3$ group are therefore substantially increased. For this spin-labeled lecithin the predominant effect of cholesterol is to increase the spacing of the lipids and allow greater molecular motion of the tempoyl group, whereas for the

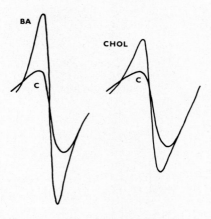

Fig. 9. Effect of (a) 80 mM benzyl alcohol, (b) cholesterol (2 lecithin:1 cholesterol) on the ESR spectrum of lecithin spin-labeled at the quaternary nitrogen (see text), in lecithin vesicles at 0 °C

unmodified lecithins observed in the NMR experiment, the effect of decreased chain motion due to cholesterol which is transmitted to the $^+NMe_3$ predominates over the effect of increased lipid spacing. For the tempoyl lecithin spin label in the presence of cholesterol these factors are approximately balanced, so that benzyl alcohol causes no net observable perturbation.

Although we have limited discussion to two prototype perturbing agents, we find a good qualitative correlation between chain motion perturbations for a range of perturbing agents and the corresponding permeability changes. It is difficult, however, to arrive at a general quantitative model relating perturbation to permeability, except for homologous series of molecules.

We can analyse this problem a little further by looking at the macroscopic parameters which determine permeation rates. The permeability of a solute can be expressed as

$$\text{Perm.} \propto P_c D$$

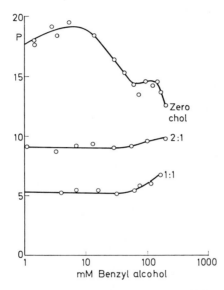

Fig. 10. The effect of cholesterol on the partition coefficient of benzyl alcohol into lecithin vesicles at 25 °C

where P_c is the partition coefficient of the solute in the bilayer and D is the diffusion coefficient in the direction perpendicular to the bilayer surface. There is an inverse relationship between P_c and D so that in general in a homologous series there will be a structure with an optimal permeability such that P_cD is maximal.

The permeability changes observed with benzyl alcohol and cholesterol should be associated with changes in one or both of these properties. The effect of cholesterol on the partition of benzyl alcohol into lecithin is shown in Fig. 10 [12]. There is a large and progressive reduction in partition with increasing cholesterol content so that at 1 lec:1 chol at low concentrations of alcohol (10 mM) the bound alcohol concentration is reduced by at least 75%, and at the critical alcohol concentration of ca. 80 mM where there is a radical change in structure induced by the alcohol, the bound concentration is reduced by at least 50%. On the other hand, benzyl alcohol has little effect either on enhancing its own partition (Fig. 10), or the partition of other small molecules such as TEMPO which is known to be a sensitive partition probe [13]. The effect of benzyl alcohol on

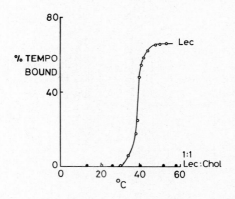

Fig. 11. The effect of temperature on the partition of 1 mM TEMPO into 230 mM dipalmitoyl lecithin, and into dipalmitoyl lecithin-cholesterol (1:1)

permeability is presumably determined mainly by an increase in the diffusion coefficient. It is probably the operation of both the partition and diffusion parameters, which are not simply related, that has made it difficult to account quantitatively for the permeability changes in terms of perturbation.

A further elegant example of the effect of chain packing on partition is provided by the thermal transition in dipalmitoyl lecithin which undergoes a crystalline to liquid crystalline transition at ca. 43 °C. The partition of TEMPO into dipalmitoyl lecithin vesicles over this temperature range increases abruptly from a very low value below the transition (Fig. 11) [3]. Cholesterol completely abolishes this partition increase [9], and the partition curve for

40 mM benzyl alcohol itself is similar to the TEMPO curve but shifted to lower temperature [12]. The perturbing agents therefore induce clear-cut changes in the thermal transition of dipalmitoyl lecithin, which is a well-defined change in the packing of the fatty acid chains in the bilayer.

Lecithin Flip Rate

KORNBERG and MCCONNELL [11] have recently used the lecithin analogue with the tempoyl group replacing a quaternary methyl

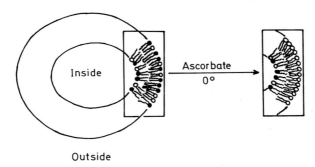

Fig. 12. The selective reduction of the exterior spin-labeled lecithin molecules in lecithin vesicles by ascorbate at 0 °C [11]

to measure the flip rate of the lecithin spin label across the bilayer. The spin label is present at ca. 5% mole fraction in the sonicated vesicles, and when ascorbate is added at 0 °C, it selectively reduces the nitroxide groups on the outside surface of the vesicles, since the ascorbate is unable to penetrate significantly at 0 °C (Fig. 12). The contribution to the spectrum of the outside facing nitroxide labels disappears, leaving only the spectrum from the nitroxide groups on the inside surface of the bilayer (Fig. 13). For the small sonicated vesicles used (ca. 250 Å diameter) approximately 70% of the nitroxide groups were exposed to the ascorbate on the outside surface of the vesicles, which was taken to reflect the difference in the number of lipids in the inner and outer surfaces of the bilayer due to the small radius of curvature. The shape of the spectrum also changes when the exterior spin labels are reduced, leaving a spectrum in which the more immobilised component is relatively enhanced (Fig. 13). This presumably indicates a difference in packing in the inner and outer lipids. To measure the flip rate, the ascorbate is removed by gel filtration, and the vesicles are allowed to equilibrate for different periods of time at 30 °C. The spin labels flip from the inside to the outside until they reach the equilibrium

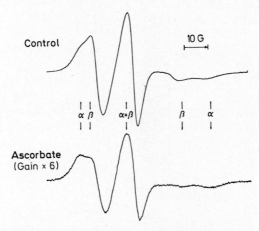

Fig. 13. ESR spectra at 0 °C of lecithin spin-labeled at the quaternary nitrogen (a) Control, (b) After reduction of exterior spin labels by ascorbate [11]

distribution and the spectrum is restored to its original shape. By following the time course of the spectral reversion, the half time for lipid flip can be determined. This was found by KORNBERG and MCCONNELL to be approximately 6 h at 30 °C (Fig. 14). Directly comparable flux data for simple ions is not available to determine whether the lipid flip rate is directly correlated with the passage of ions across the bilayer.

Not surprisingly, benzyl alcohol and cholesterol have opposite effects on the flip rate [9]. Benzyl alcohol at 30 mM approximately doubles the flip rate at 25 °C and the half time drops to about 3 h (Fig. 15). In the presence of equimolar cholesterol, the flip rate at

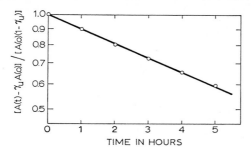

Fig. 14. Time course of the flip of spin-labeled lecithin molecules from inside the lecithin vesicle to the outside surface. The abscissa is the calculated fraction of the spin-labeled molecules which have flipped [11]

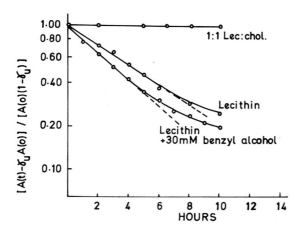

Fig. 15. The effects of benzyl alcohol and cholesterol on the flip rate of spin-labeled lecithin in lecithin vesicles at 25 °C

25 °C was too slow to measure over 8 h. These are preliminary results but they reflect the qualitative effects of alcohol and cholesterol on the flip rate.

It is worth commenting on the relationship of these passive permeability changes induced by the perturbing agents in lecithin vesicles to the membrane effects of benzyl alcohol and cholesterol. A number of diverse membrane effects of these agents are shown in

Table 3. The important point to be emphasized is that in the lipid vesicles, the permeability changes and the correlation with dynamic perturbations is well established, and it is to be expected that these agents will cause similar qualitative changes in passive permeability of the lipid part of the structure in biological membranes. This passive leak effect must not be confused, however, with the effects of perturbing agents on facilitated diffusion which is protein mediated. An example in Table 3 from the work of MARTIN [14] illustrates the point clearly; 40 mM benzyl alcohol causes a 70% decrease in the facilitated diffusion of choline into the erythrocytes, and there

Table 3. *Membrane effects of benzyl alcohol and cholesterol*

System	Benzyl alcohol	Cholesterol
Lecithin vesicles	↑ permeability	↓ permeability
Erythrocytes	↓ facilitated diffusion of choline and glucose[14]	—
Monkey kidney cells in primary culture	↓ Growth rate 50% decrease at 40 mM[15]	↓ Growth rate
Frog sciatic nerve	Minimum blocking conc^n = 36 mM[16]	—

is a similar inhibition of facilitated glucose transport at the same alcohol concentrations.

These are clearly effects of the alcohol on protein functions quite distinct from, and in the opposite direction to, the effect on passive "leak" through the membrane. There is no evidence at present whether the effect of an agent such as benzyl alcohol is due to direct interaction of the alcohol with the protein carrier, or an indirect action through perturbation of the lipids which seal the carriers into the membrane structure.

It is this problem of observing lipids directly interacting with membrane proteins which we are now trying to tackle. It is clear from the preceding discussion that the complexity of the structure of biological membranes by itself presents a formidable analytical problem using any technique capable of solving the problem at the molecular level. We conclude that, as in the simple lipid vesicles, to obtain useful NMR spectra from membranes will require simplification by specific isotopic substitution to detect and identify each chemical group of interest in the structure in turn. The problem is illustrated in Fig. 16, which shows one of the more promising NMR spectra from unsonicated and unmodified membranes. The ^{13}C NMR spectrum of the erythrocyte membrane [1], although much more detailed than the corresponding proton spectrum which contains

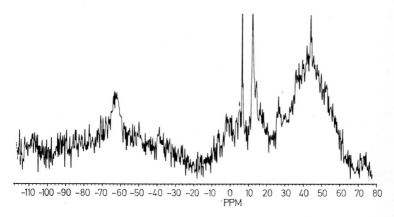

Fig. 16. The ^{13}C NMR spectrum of erythrocyte membranes (10% w/w) at 28 °C

no high resolution detail, indicates that both enrichment to improve the signal to noise ratio and selective substitution to simplify the spectrum are essential for further progress.

Probe Experiments with Membranes

It is much easier technically to observe molecules with suitable NMR spectra as empirical probes, than to attempt to observe labeled membrane components directly. Information about the organization of the membrane structure is inferred indirectly from its interaction with the probe. Although probes can provide a sensitive measure of change in the structure in response to perturbation, we do not usually know precisely where the probe is localised in the membrane, so that there are obvious limitations on the structural information it can provide. However, in spite of these limitations, which are common to all the spectroscopic experiments of this kind, we find that we can readily distinguish the main types of binding sites for probes within the membrane by their spectroscopic properties, and use these binding sites to define structural features of the membrane.

The technique is illustrated for the same perturbing agent, benzyl alcohol, used previously in the lipid system. In the presence of an erythrocyte membrane suspension, the resonance of the aromatic ring of benzyl alcohol is broadened by stearic interaction of the bound alcohol molecules with the membrane [17]. The observed resonance linewidth ($\Delta v_{1/2}$) is the weighted mean of the very broad signal from the bound molecules and the sharp signal from the large excess of free alcohol molecules in solution. The observed linewidth changes as the alcohol concentration in the membrane is increased (Fig. 17); the probe itself is used to report on the perturbation it causes in the structure. At low alcohol concentrations the linewidth is relatively broad, but becomes narrower at higher concentrations, so that the alcohol molecules find themselves in a more fluid environment in the membrane. This continues until, at a critical concentration of about 80 mM, the linewidth increases sharply again. At this critical concentration, benzyl alcohol also causes lysis of intact erythrocytes. Up to the lytic concentration the linewidth changes are fully reversible on removing the alcohol, but in the lytic concentration range the broadening is irreversible. On reducing the concentration from about 200 mM benzyl alcohol, the linewidth in the presence of these pretreated membranes increases monotonically, and is much broader than for the original membranes (Fig. 17). Clearly the interaction of the alcohol with the disrupted membrane is drastically altered, and this is also associated with an increase in the number of binding sites available to the alcohol by a factor of about 2 compared with the intact membrane [18].

It is readily shown that most of these abnormal binding sites are localized on the membrane protein by repeating the linewidth measurements in the presence of the separated membrane lipids and proteins. The lipid is reconstituted in the form of vesicles and causes very little broadening of the resonance over the whole concentration range (Fig. 17). The protein ($> 95\%$ of all the mem-

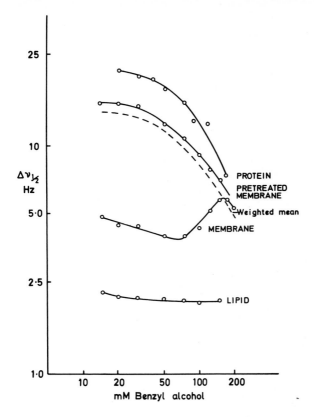

Fig. 17. The effect of erythrocyte membranes and various derived membrane preparations on the linewidth ($\Delta v_{1/2}$) of the aromatic protons of benzyl alcohol at 25 °C (see text). All preparations are 1.0% by weight. The dashed curve is the mean linewidth of the separated lipid and protein components corresponding to the composition of the membrane [33]

brane proteins) is in a soluble form and causes greatly enhanced broadening over the whole concentration range. It is also clear that the broadening is much greater than can be accommodated in the intact membrane. If we calculate the broadening from the separated lipid and protein curves corresponding to the composition of the intact membrane, it follows the dashed curve in Fig. 17. It is much greater than the broadening of the intact membrane over the low concentration range and is obviously dominated by the contribution from the membrane protein.

Our conclusions from this experiment [17, 19, 20] are that at low concentrations the alcohol progressively fluidises the membrane structure until the lytic concentration is reached. In this low concentration range many of the binding sites on the separated membrane components are unavailable to the alcohol, through the assembly of the components into the intact membrane structure. The alcohol molecules only bind to sites which are not essential to

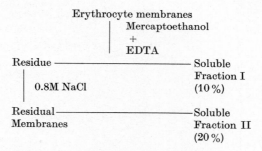

Fig. 18. Fractionation of erythrocyte membrane proteins by mild aqueous washing solutions. (Adapted from ROSENBERG and GUIDOTTI [23])

the integrity of the membrane structure, and the interaction is fully reversible. Above the lytic concentration, the alcohol molecules progressively compete for essential interaction sites between the membrane components, exposing an abnormal set of binding sites which are not accessible in the intact structure. The majority of these abnormal sites are clearly localized on the membrane protein. At maximal alcohol concentrations (200—300 mM) the fully disrupted membrane has nearly all the binding sites exposed which are available in the separated components, so that it interacts with the alcohol essentially as the sum of its separated components. These conclusions have been confirmed in detail by alcohol binding studies [18] and by independent probe experiments with spin labels [21] and fluorescence probes [22], which readily detect the abnormal protein binding sites and can be used as assays.

Using this simple NMR linewidth measurement as a criterion, we can ask more detailed questions about the protein binding sites. For example, do all the membrane proteins carry these abnormal binding sites or are they limited to a distinct group of proteins which may be regarded as essential for the integrity of the structure? Mild aqueous washing procedure due to ROSENBERG and GUIDOTTI [23] (Fig. 18) removes between 30% and 50% of the erythrocyte

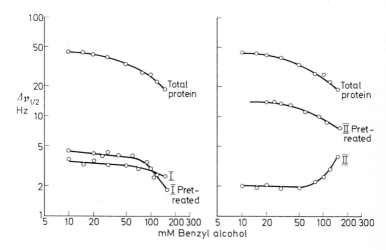

Fig. 19. The linewidths ($\Delta v_{1/2}$) of the aromatic protons of benzyl alcohol in the presence of erythrocyte membrane protein fractions I and II at 25 °C. The fractions are compared with their linewidths after treatment with 300 mM benzyl alcohol and with the complete membrane protein fraction

membrane proteins into solution, and the presence of abnormal binding sites on these separated soluble proteins is assayed by measuring the linewidth of the alcohol phenyl resonance in the presence of the proteins. Fig. 19 shows clearly that neither fraction contains a significant proportion of the binding sites when separated [24]. On pretreating the two fractions with 250 mM benzyl alcohol, which would expose the full complement of any latent binding sites, there is no significant change in the linewidth from fraction I, but a significant increase in fraction II. Even after pretreatment, however, the linewidth from fraction II is only of the same order as the native membrane, and only 30% of the linewidth of the total population of membrane proteins, so that it is relatively poor in abnormal binding sites. The experiment also indicates that the residual membranes remain structurally intact after removal of fractions I and II, as judged by the biphasic curve

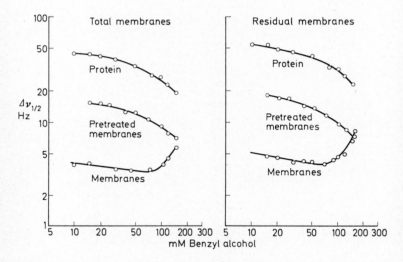

Fig. 20. As Fig. 19, comparing the total membrane linewidth curve at 25 °C with the residual membrane curve after removal of fractions I and II. Both membrane curves are compared with the complete membrane protein fraction and with the curves obtained after pretreating the membranes with 300 mM benzyl alcohol. All preparations are 1.0% w/w

from the residual membranes (Fig. 20). This is a characteristic feature of a wide range of intact membrane [25] structures and provides a stringent test of the state of the membrane structure [20], illustrated in the following experiments.

If it were possible to separate the membrane into its lipid and protein components, label the components with appropriate groups for spectroscopy, and then reconstitute the membrane into its native structure, detailed information could be obtained about the interactions between membrane components. Although the correct reassembly of so many components seems unlikely there are several reports that membranes which have been substantially separated into lipid and protein fractions can be reconstituted to structures resembling the original membrane. The most detailed studies have been of reaggregated membranes from Mycoplasma laidlawii [26 to 30]. The original membranes are dissolved in sodium dodecyl sulphate (SDS) solution to form soluble complexes of detergent with the lipids and proteins which can be separated by gel filtration. On removing the SDS by dialysis under appropriate conditions against a buffer containing Mg^{2+} ions, reaggregated structures are formed which resemble the original membranes by several criteria. The reaggregate contain all the lipid and protein of the original membranes and also have the same buoyant density of 1.18 which is between the densities of the separated lipid and protein components. The characteristic triple-layered structure of the original membranes in electronmicrographs is retained by the reaggregated structures. The thermal transition in the lipid bilayer of the membrane [31], in which X-ray diffraction measurements show that the lateral spacing of the lipid chains increases from 4.15 to 4.6 A [32] on increasing the temperature through the transition, is restored in the reaggregates [33]. This indicates that a lipid bilayer is substantially reformed in the reaggregates, but does not give any information about the state of the membrane proteins.

The original Mycoplasma membranes show a biphasic benzyl alcohol curve characteristic of a native membrane [33], but it is clear that the reaggregates are intermediate between the fully separated components, which coincide with the pretreated membrane curve, and the intact membrane (Fig. 21). The best reaggregates to date show a small inflection at the critical concentration but they lack the full biphasic response of the intact membrane. In particular, many of the protein binding sites which should be inaccessible in the intact membrane are still available in the reaggregates.

This approach does not therefore allow the facility of specific labeling of membrane components which we require. The most important conclusion from these empirical probe studies is that the binding of probe molecules is determined primarily by the organization of the membrane components which justifies the use of probes as limited structural determinants. The full exploitation of the NMR technique now rests on the direct observation of specifically labeled membrane components.

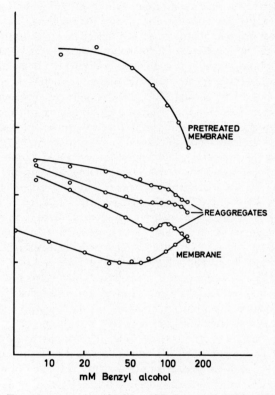

Fig. 21. As Fig. 19, comparing the linewidth curves at 25 °C of Mycoplasma membranes with the reaggregated membrane structures and membranes pretreated with 300 mM benzyl alcohol. All preparations are 1.0% w/w

References

1. METCALFE, J. C., BIRDSALL, N. J. M., FEENEY, J., LEE, A. G., LEVINE, Y. K., PARTINGTON, P.: Nature (Lond.) **233**, 199 (1971).
2. BIRDSALL, N. J. M., LEE, A. G., LEVINE, Y. K., METCALFE, J. C.: Biochim. biophys. Acta (Amst.) **241**, 693 (1971).

3. HUBBELL, W. L., MCCONNELL, H. M.: J. Amer. Chem. Soc. **93**, 314 (1971).
4. MCFARLAND, B. G., MCCONNELL, H. M.: Proc. nat. Acad. Sci. (Wash.) **68**, 1274 (1971).
5. HUBBELL, W. L., MCCONNELL, H. M.: Proc. nat. Acad. Sci. (Wash.) **64**, 20 (1969).
6. LEVINE, Y. K., WILKINS, M. H. F.: Nature (Lond.) **230**, 69 (1971).
7. DE GIER, J., MANDERSLOOT, J. G., VAN DEENEN, L. L.: Biochim. biophys. Acta (Amst.) **150**, 660 (1968).
8. KLINE, R. A., MOORE, M. J., SMITH, M. W.: Biochim. biophys. Acta (Amst.) **233**, 420 (1971).
9. HOULT, J. R. S., METCALFE, J. C.: Unpublished data.
10. LEE, A. G., LEVINE, Y. K., BIRDSALL, N. J. M., METCALFE, J. C.: Biochim. biophys. Acta (Amst.) (1971) (in press).
11. KORNBERG, R. D., MCCONNELL, H. M.: Biochemistry **10**, 1111 (1971).
12. COLLEY, C. M., METCALFE, J. C.: Unpublished data.
13. HUBBELL, W. L., MCCONNELL, H. M.: Proc. nat. Acad. Sci. (Wash.) **61**, 12 (1968).
14. MARTIN, K.: J. Physiol. (Lond.) (1971) (in press).
15. METCALFE, S. M.: J. Pharm. Pharmacol. (1971) (in press).
16. SCHAUF, C., AGIN, D.: Nature (Lond.) **221**, 768 (1969).
17. METCALFE, J. C., SEEMAN, P. M., BURGEN, A. S. V.: Molec. Pharmacol. **4**, 87 (1968).
18. COLLEY, C. M., METCALFE, S. M., TURNER, B., BURGEN, A. S. V., METCALFE, J. C.: Biochim. biophys. Acta (Amst.) **233**, 720 (1971).
19. METCALFE, J. C.: In: Calcium and cellular function, p. 219 (CUTHBERT, A. W., Ed.). London: Macmillan 1970.
20. — In: Permeability and function of biological membranes, p. 222 (BOLIS, LIANA, KATCHALSKY, A., KEYNES, R. D., LOEWENSTEIN, W. R., PETHICA, B. A., Eds.). Amsterdam: North-Holland Publishing Co. 1970.
21. HUBBELL, W. L., METCALFE, J. C., METCALFE, S. M., MCCONNELL, H. M.: Biochim. biophys. Acta (Amst.) **219**, 415 (1970).
22. METCALFE, S. M., METCALFE, J. C., ENGELMAN, D. M.: Biochim. biophys. Acta (Amst.) **241**, 422 (1971).
23. ROSENBERG, S. A., GUIDOTTI, G.: J. biol. Chem. **244**, 5118 (1969).
24. RANDALL, R. F., STODDART, R. W., METCALFE, S. M., METCALFE, J. C.: Biochim. biophys. Acta (Amst.) (1971) (in press).
25. METCALFE, J. C., BURGEN, A. S. V.: Nature (Lond.) **220**, 587 (1968).
26. ENGELMAN, D. M., TERRY, T. M., MOROWITZ, H. J.: Biochim. biophys. Acta (Amst.) **135**, 381 (1967).
27. TERRY, T. M., ENGELMAN, D. M., MOROWITZ, H. J.: Biochim. biophys. Acta (Amst.) **135**, 391 (1967).
28. RAZIN, S., MOROWITZ, H. J., TERRY, T. M.: Proc. nat. Acad. Sci. (Wash.) **54**, 619 (1965).
29. — NE'EEMAN, Z., OHAD, I.: Biochim. biophys. Acta (Amst.) **193**, 277 (1969).
30. — KAHN, I.: Nature (Lond.) **223**, 863 (1969).

31. Stein, J. M., Tourtellotte, M. E., Reinert, J. C., McElhaney, R. N., Rader, R. C.: Proc. nat. Acad. Sci. (Wash.) **63**, 104 (1969).
32. Engelman, D. M.: J. molec. Biol. **47**, 115 (1970).
33. Metcalfe, J. C., Metcalfe, S. M., Engelman, D. M.: Biochim. biophys. Acta (Amst.) **241**, 412 (1971).

Synthetic Lipid- and Lipoprotein Membranes

Hans Kuhn

Max-Planck-Institut für biophysikalische Chemie,
Karl-Friedrich-Bonhoeffer-Institut, Göttingen

By superimposing monomolecular layers it is possible to arrange molecules in a specific planned order. This method of obtaining simple organized systems of molecules should be of interest to the membranologist as a possible means of building models of biological structures. The possibility of molecular contact between a synthetic monolayer assembly and a biological membrane may permit new types of interference between an organized system of monolayers and the biological machinery.

Models of organized layer systems can be built by depositing layers of dyes and fatty acids on suitable supports [1]. Mixed layers of fatty acids and of cyanine dyes containing a hydrophilic chromophore with long paraffin substituents are particularly suitable, since the paraffin substituents are firmly incorporated into the monolayer of the fatty acid [2]. The energy transfer between layers of different species of dyes separated by monolayers of fatty acids has been studied by measuring the fluorescence quenching of the sensitizer dye or the sensitized fluorescence of the energy acceptor dye [1, 2, 3]. The thickness of a biological membrane is found by sandwiching it between layers of sensitizer and acceptor, and measuring the amount of energy transfer across the membrane [4].

The amount of energy transfer between the layers of the sensitizer dye and the acceptor dye depends strongly on the distribution of the acceptor dye within the layer [2]. In the usual case of a mixed film of dye and fatty acid the dye molecules are statistically distributed within the monolayer. In some cases, however, no mixed films are obtained, but mosaic structures of clusters of pure dye and of pure arachidic acid. The fluorescence of the sensitizer is much less quenched by the acceptor layer with the mosaic structure, since a large portion of the number of sensitizer molecules is not in the close vicinity of the acceptors. The amount of quenching is determined by the size of the clusters of acceptor dye molecules [5]. The technique for estimating the distribution of the acceptor dye by energy transfer measurements may be useful to the membranologist for localizing natural chromophores or artificial dyes incorporated in the biological membrane. The dye or chromophore may act as the acceptor in an assembly where the sensitizer is built in a monolayer in contact with the membrane.

The monolayer assembly technique can be used for investigating the deactivation of excited molecules and the spectral sensitization [6]. It is found that a fluorescent dye molecule is able to sensitize the photographic process at distances up to 300 Å from the surface of a silver bromide crystal [7]. This again might indicate a possible means of locating dyes in a biological membrane: the membrane might be placed in contact with the silver bromide surface. After exposure to light which is absorbed by the dye, followed by development, the sensitized photographic picture should indicate the distribution of the sensitizing dye within the membrane. Small molecules can diffuse readily through the synthetic layer assemblies, whereas larger molecules do not penetrate into the layer system but adhere to the surface of the system. By the energy transfer method it is easy to establish whether or not diffusion through the layer system takes place [8]. The stabilities of the layer assemblies and their permeability for small molecules would be important for the possible use of such systems as models of biological structures or as active components built onto biological membranes.

Lipid-protein layer systems can be obtained by adsorption of proteins from the subphase onto a lipid layer [9]. Penetration can be avoided by application of a sufficiently high surface pressure to the film onto which the protein is being adsorbed. Organized assemblies containing such layers can be constructed and their architecture can be controlled by energy transfer techniques. Thus, it seems possible to build functional units containing several cooperating enzymes.

A molecular monolayer can be precisely removed from a suitable support. Further, it is possible to divide a system of suitable structure into precisely determined layers. The monolayer system can be manipulated in such a way that it can be attached on either side to another monomolecular layer. The process of making molecular contact can be checked readily at the various stages by the energy transfer method [10]. New possibilities for monolayer construction are offered by this procedure.

Another method of bringing layers into molecular contact and of separating them again is by making contact between two soap bubbles and subsequently separating them [11]; different fluorescent surface-active dyes are incorporated into the two soap lamellae. It is found that no rearrangement of the layers and no diffusion of dye molecules from one lamella into the other take place during the few seconds contact time. Thus, the soap lamellae in contact with each other have a double-lamella structure. The experiments described offer a new method for producing models of bimolecular lipid membranes.

Cyanine dyes can be incorporated with pre-determined geometry into a layer arrangement [12]. The dye monolayer can be sandwiched between monolayers of cadmium arachidate held between semitransparent metal electrodes [13]. The monolayer condenser withstands electric fields up to $5 \cdot 10^6$ V/cm with an applied voltage of 10 V. The electric field-induced shift of the absorption band can be determined by measuring the change in transmittance of the monolayer condenser as a function of the applied voltage. Signs and approximate values of the shifts are estimated by an electron-gas model calculation [13]. This quantitative approach is of interest in connection with the use of dyes in biological membranes as indicators for the electric field in the membrane [14].

Capacitors consisting of two metal layers and a thin insulating monolayer can be produced by the monolayer assembly technique [15]. A potential of, e.g., 1 V is applied and the current is measured. The thickness of the insulating monolayer can be varied in small steps by proceeding from one fatty acid to another with a different C-atom chain length. The current decreases exponentially with the increasing of this thickness. This current is due to the quantum-mechanical tunneling of electrons through the insulating layer. The dependence of the current on voltage and layer thickness, as well as on the effects due to the electrode metal, can be quantitatively interpreted by tunneling theory. The excellent agreement between theory and experiment is an extremely sensitive test on the uniformity of the monomolecular layer. Holes and conducting impurities in the layer would strongly influence the electric resistance.

The photo-current of an arrangement consisting of a monolayer of a dye sandwiched between fatty acid monolayers held between semitransparent electrodes increases exponentially with the applied voltage. It also depends on the position of the dye monolayer within the assembly and shows an action spectrum identical to the absorption spectrum of the dye monolayer [16]. The model, which leads to a quantitative description of these results, is based on the assumption that the electron in the optically excited level of the dye molecule may be thermally excited to the top of the energy barrier of the arachidate and then moved to the positively biased metal. By applying a voltage the energy barrier is diminished by an amount which depends on the length of the potential trough substituting the dye molecule. The length of the trough was calculated and it was found that this length agrees with the experimental length of the dye molecule.

The good agreement between theory and experiment indicates that the electronic processes in such monolayer assemblies are well understood and that it should be possible to control such processes. For example an appropriate potential profile for a light-driven electron pump may be determined and then this profile might be constructed by assembly of the known component monolayers. In this way the monolayer assembly technique could contribute towards obtaining synthetic models of interest in molecular biology and in molecular engineering.

References

1. ZWICK, M. M., KUHN, H.: Z. Naturforsch. **17a**, 411 (1962).
   DREXHAGE, K. H., ZWICK, M. M., KUHN, H.: Ber. Bunsenges. Phys. Chem. **67**, 62 (1963).
   BARTH, P., BECK, K. H., DREXHAGE, K. H., KUHN, H., MÖBIUS, D., MOLZAHN, D., RÖLLIG, K., SCHÄFER, F. P., SPERLING, W., ZWICK, M. M.: In: Optische Anregung organischer Systeme, p. 639. Weinheim: Verlag Chemie 1966.
   KUHN, H.: Pure App. Chem. **11**, 345 (1965).
   — Naturwissenschaften **54**, 429 (1967).
   — In RICH, A., DAVIDSON, N.: Structural chemistry and molecular biology, p. 566. San Francisco and London: W. M. Freeman 1968.

2. BÜCHER, H., DREXHAGE, K. H., FLECK, M., KUHN, H., MÖBIUS, D., SCHÄFER, F. P., SONDERMANN, J., SPERLING, W., TILLMAN, P., WIEGAND, J.: Molec. Cryst. **2**, 199 (1967).
3. MÖBIUS, D.: Z. Naturforsch. **24a**, 251 (1969).
4. PETERS, R.: Biochim. biophys. Acta (Amst.) **233**, 465 (1971).
5. BÜCHER, H., v. ELSNER, O., MÖBIUS, D., TILLMANN, P., WIEGAND, J.: Z. physik. Chem. (Frankfurt) NF **65**, 152 (1969).
6. INACKER, O., KUHN, H., BÜCHER, H., MEYER, H., TEWS, K. H.: Chem. Phys. Lett. **7**, 213 (1970).
7. v. SZENTPÁLY, L., MÖBIUS, D., KUHN, H.: J. chem. Phys. **52**, 4618 (1970).
8. KUHN, H., MÖBIUS, D.: Angew. Chem. **83**, 672 (1971); Internat. Ed., **10**, 620 (1971).
9. FROMHERZ, P.: Biochim. biophys. Acta (Amst.) **225**, 382 (1971).
  Nature (Lond.) **231**, 267 (1971).
  FEBS Lett. **11**, 205 (1970).
10. MÖBIUS, D., INACKER, O., KUHN, H., to be published, see 8.
11. MORMANN, W., KUHN, H.: Z. Naturforsch. **24b**, 1340 (1969).
12. BÜCHER, H., KUHN, H.: Chem. Phys. Lett. **6**, 183 (1970).
13. BÜCHER, H., KUHN, H.: Z. Naturforsch. **25b**, 1323 (1970).
  — WIEGAND, J., SNAVELY, B. B., BECK, K. H., KUHN, H.: Chem. Phys. Lett. **7**, 508 (1969).
  KLEUSER, D., BÜCHER, H.: Z. Naturforsch. **24b**, 1317 (1969).
14. WITT, H. T.: Z. Naturforsch. **236**, 244 (1968).
  — RUMBERG, B., JUNGE, W., DÖRING, G., STIEHL, H. H., WEIKARD, J., WOLFF, CH.: Progr. in Photosynthesis Res., **3**, 1361 (1969).
  JUNGE, W.: Europ. J. Biochem. **14**, 582 (1970).
15. MANN, B., KUHN, H.: J. appl. Phys., scheduled to appear in Nov. 1971 issue.
  — — v. SZENTPÁLY, L.: Chem. Phys. Lett. **8**, 82 (1971).
16. SCHOELER, U., KUHN, H., BÄSSLER, H., TEWS, K. H.: To be published.

Round Table Discussion

SCHMITT: Ladies and Gentleman, the program committee has given me the formidable task of organizing this "round table"; I have accordingly suggested a list of topics in order to pull together the major issues raised during the two days of plenary discussions. As I have listed them on the board they are ***self-assembly, biosynthesis, passive permeation, cell-cell recognition in differentiation, membrane constitution, cooperativity in membranes, transformation,*** and ***receptor theory.*** Almost nothing has been said about active transport, which is unusual since the transport "union" is very strong. Perhaps we will have time to touch on this topic.

Now, without further ado, I will call on certain individuals to start discussions and hope that others will participate. But time is precious and I must ask you to focus your remarks as much as possible.

Starting with *self-assembly*, let us raise an issue that Dr. OVERATH brought up yesterday and ask him and Dr. STOFFEL to deal with the question in the interrelationship of lipid and protein biosynthesis. Dr. OVERATH suggested that these processes are independent, but Dr. ROTHFIELD on the other hand showed that if we are going to reconstitute or constitute an enzymatic reactive lipoprotein system, we must first have the constituents synthesized.

Self-Assembly and Biosynthesis

OVERATH: *First*, it is my view that, if there is an enzyme complex in the membrane which makes lipid, lipid synthesis in the membrane is independent of protein synthesis at a ribosome. *Secondly*, there is no evidence at the moment that various ribosomes differ in their functional potential, although the question of ribosome structure and their functional differentiation within cells is very much in the open. *Finally*, if we assume that ribosomes differ functionally, e.g. free versus membrane-bound ribosomes, the corresponding messengers would require recognition regions, guiding them to the appropriate free or membrane-bound ribosomes, but there are no data for or against this notion.

SCHMITT: That clarifies the point for me. Dr. STOFFEL, would you care to comment? OVERATH has stressed regulation at the level of the ribosome, but there must also be later control points; I am thinking of lipid synthesis.

STOFFEL: The question whether the synthesis of membrane lipids is completely unrelated to membrane protein synthesis, as Dr. OVERATH postulates, or whether there is some kind of regulation between the two cannot be answered at the present time since no experimental evidence is available.

It is, however, striking to observe the constancy of the lipid content and the composition of the constituents of the different membranes in animal cells apparently related to the function of these membranes. This suggests some regulation between the two membrane components. Preliminary studies carried out by Dr. SCHIEFER in my laboratory suggest that lipid synthesis in rat liver cells declines approximately 10 min after the cessation of protein synthesis blocked with chloramphenicol.

LYNEN: I really do not understand Prof. SCHMITT's question. We are discussing two distinct problems. Dr. OVERATH has discussed synthetic mechanisms in membranes, whereas ROTHFIELD has reconstituted a membrane system for a lipophilic substrate. His acceptors for galactose or glucose are distinctly lipophilic; hence three components are required — the phospholipid, the enzyme and the substrate.

SCHMITT: I agree, and what has just been discussed clarifies the matter for me. OVERATH dealt with the site of synthesis, but there was some misunderstanding as to whether the lipid and protein work closely together in the building and reconstitution of membranes. I am sorry that Dr. ROTHFIELD could not stay to comment on this and tell us something about phage assembly. The manner in which macromolecules assemble themselves without direct genetic control is something I consider very important for membrane biology. Perhaps Dr. TRENT could tell us something about a related problem, the assembly of vaccinia virus.

TRENT: This is really not my field, but I can outline the process. I am not speaking of viruses assembled in such a fashion that the viral membranes form in continuity with pre-existing cellular membranes. We do not know whether the pre-existing membranes act as a framework where virus specific substances replace pre-existing ones in stepwise fashion, or whether they act as crystallization points for virus products, which then displace normal membrane components. The vaccinia viruses are favorable in such studies because of their large size and a complexity approaching that of bacteria. In infected cells one early finds membrane areas which appear typical for that virus. Somewhat later a very peculiar phenomenon develops, i.e. small membrane pieces, forming at the edges of the virus-specific patches, with the appearance of "unit membranes", but with free, slightly curved ends, and a rather stiff center portion. The real, inner virus components develop later within these "half-moons". They are rather complicated, with a central moiety and two, so-called nucleotides, all of which are ultimately enveloped by the membrane. The completed virus particles are released when the cell bursts. The independent virus membranes were never continuous with normal cellular membranes, as is also apparent from their lipid composition, which is quite distinct from that of adjacent cellular membranes, e.g. endoplasmic reticulum.

SCHMITT: Thank you very much. Certainly self-assembly and the possible role of specific induction of genetic factors in a sequential fashion are extremely important to membranology. Does anyone care to add to that? Please, Dr. HESS:

HESS: I would like to formulate the question by asking whether there is a compulsory order of incorporation of proteins into membranes. I think a good example is the induction of mitochondrial organization from the promitochondrial state, in which case there seems to be no such order. The promitochondria contain quite a number of enzymes which are active in the respiratory chain, such as oligomycin-sensitive ATPase and particle-bound dehydrogenase. Upon induction with very small concentrations of oxygen cytochrome develops and can be isolated in particulate form. If one repeats the experiment in the presence of chloroamphenicol, cytochrome-C is synthesized. Induction with air results in an assembly of the respiratory chain. It seems that by varying the induction conditions one can change the ratios between the cytochromes relative to labeled protein or ATPase, and from this one could conclude that, at least in yeast, there is no compulsory order of protein insertion. This might, of course, be completely different in higher organisms.

SCHMITT: Dr. LEHNINGER, would you like to come in on that? If you recall, I indicated in my introduction that the lipid pattern in each cell is probably unique, and Dr. LEHNINGER suggested some 10 years ago that the pattern of polar groups at the inner and outer lipid interfaces might be a code, determining the type of proteins and other compounds that could bind there. Would you please state your present views on that point?

LEHNINGER: First, let me deal with a question brought up earlier and which Dr. HESS has commented on. It is true that the self-assembly of some supramolecular systems may not involve any specific order of assembly, but in the case of T4 bacteriophage assembly a definite sequence does occur, as shown by WOOD and his colleagues. Each of the parts is assembled seperately and in a given sequence; the head, tail, and tail fibers then combine, again in a specific order. Moreover, at a membrane symposium at Oak Ridge two weeks ago, STROMINGER and ROSEMAN both reported that lipid-requiring enzymes are necessary to construct different components of the bacterial cell wall; also, in these cases a specific order of addition of lipid to protein was absolutely essential for reconstitution.

The other matter Dr. SCHMITT touched on was the specificity of lipids in different types of membranes. There are certain specific lipids, which are characteristically present in one or another type of membrane. For example, the lipids in the inner membrane of rat liver mitochondria contain over 20% cardiolipin, together with a distribution of other phospholipids. In heart mitochondria one again finds cardiolipin as a major component, although the proportion of the other lipids differs. Cardiolipin is a major component of the inner mitochondrial membrane in all species so far examined, but the other lipids in the membrane vary widely in composition. Cardiolipin is thus organelle-specific. On the other hand, if one compares the membrane lipids of mitochondria with those of endoplasmic reticulum in the same cell one generally finds certain similarities in lipid distribution, that is, cell or tissue specificity. So membrane lipid patterns show organelle specificity, tissue specificity and species specificity. The phospholipid distribution in each type of animal membrane is relatively constant and probably genetically fixed: it can not be altered by simple nutritional changes. The important question is what determines the phospholipid distribution in each biomembrane. Is it the protein composition, i.e. do membrane proteins code the specific phospholipid distribution of a membrane or does the phospholipid composition determine the protein composition? I suspect that it is probably the membrane proteins which determine the phospholipids. If that is the case, one has to look for some kind of specific interaction between membrane proteins and lipids. According to this hypothesis, the membrane proteins, whether catalytic or not, must contain specific binding sites that could accept the polar heads of specific phospholipids, and ensure in this way that the specific ratio of the different sorts of lipids characteristic for that membrane is formed as the membrane undergoes synthesis. This would then suggest that each protein molecule in the membrane has one or more specific sites, specific for certain lipids, analogous to the substrate binding sites of enzymes. We have tried to examine this proposition, with some discouraging results. The entry to this problem came from the interesting work by van Deenen and associates, who showed that the permeability of erythrocytes from different species varies with the phospholipid composition of the erythrocyte membrane, which in turn differs widely from one mammalian species to another. These workers also carried out gel-electrophoresis of the erythrocyte membrane proteins from these different species, with the ultimate hope that the different kinds of proteins in the membranes of different species of erythrocytes stand in some kind of correspondence with the type and characteristic distribution of the phospholipids. We have tried to isolate the major protein species from erythrocyte membranes of different species and then through various procedures, which I won't detail, to see whether the different proteins from these membranes had different and characteristic lipid-binding specificities. Unfortunately, every membrane protein we isolated, regardless of species, bound all phospholipids to an equal extent. Of course, when one unfolds membrane proteins with detergents or organic solvents they cannot necessarily refold again to regenerate specific binding sites for lipids. Despite the failure of this approach to date, for obvious technical reasons, I believe that it will be necessary to test this hypothesis as soon as the technique of renaturing insoluble membrane proteins has been better developed.

In bacteria the situation is somewhat different because one can vary their membrane phospholipid distribution by very simple procedures. If one grows *E. coli* cells in media of wide pH, their phospholipid composition changes. Usually the acidic lipids are diminished when the cells are grown in acid media. Obviously certain *external* influences can alter the lipid composition of *E. coli* membrane. Thus I think we are left with some basic and fundamental problems concerning the regulation and the genetic determination of the synthesis and assembly of both membrane proteins and lipids. I am also inclined to agree with Dr. OVERATH that the regulation of these two biosynthetic processes probably takes place at different sites in the cell.

SCHMITT: As I mentioned in my description of myelination, early myelin ("premyelin") contains Schwann or glial cell membranes only loosely wrapped. To form mature, densely packed myelin the synthesis and incorporation of long galactolipid, cerebroside molecules is required. I would suggest that these molecules may play an important role in effecting cell adhesion generally.

WALLACH: Concerning lipid oxidation, one should recall that this is a function not only of the number of unsaturated double bonds but also of the local pO_2, particularly where lipid synthesis occurs and where most cells, other than alveolar macrophages, reside is extremely low, on the average of less than 30 mm Hg rather than the 100 mm Hg in the alveoli. We should also recall that the work of MCCONNELL and METCALFE, using spin labels, showed that the fluidity of the membrane, particularly near its middle, varies with the number of double bonds and that the presence of unsaturated groups certainly produces pliancy in these chains; that must have an important physical aspect.

SCHMITT: We have spoken about passive permeation, but transport has scarcely been mentioned. Does someone want to speak about this topic?

Passive Permeation and Transport

PASSOW: Since many other aspects of membrane biology have already been covered, I wish to follow Prof. SCHMITT's suggestion and focus on membrane transport. I should like to present some experiments on the effects of lead on the potassium permeability of erythrocyte membranes. These experiments may serve to illustrate some of the problems which arise if one attempts to interpret kinetic measurements on cell suspensions.

The addition of 10^{-7} moles of lead acetate/gr of erythrocytes causes a thousandfold increase in K efflux. K^+ loss occurs in two phases. The first is very rapid, the second is slow. The transition from the first to the second phase always occurs at about the same time after the start of the experiment by the addition of lead. For an interpretation of this peculiar time course, we considered the following alternative: (1) at the end of the first phase of rapid loss, all cells spontaneously recover for the lead effect, or else (2) lead gives rise to an all-or-none effect. A given cell either leaks or it doesn't. After the leaky cells have lost all of their potassium at the rapid rate observed during the initial phase of the experiment, the remaining ones still contain nearly their normal complement of intracellular K. This K^+ leaves the cells at the much slower rate observed during the second phase of K loss.

Lead only increases K efflux but not Na influx. Hence, lead poisoning causes shrinkage of the cells. This can be detected by means of the Coulter Counter. Measurements of the volume distribution of the cell population with this instrument show that, with increasing lead concentration, an increasingly larger fraction of the cells of a population shrinks whereas the remaining cells retain nearly their original volume. One can demonstrate that the shrunken cells contain much less potassium than the normal cells. Hence, the data point to an all-or-none phenomenon.

Nevertheless, spontaneous recovery of leaky cells from the effects of lead can also be demonstrated. This spontaneous recovery first becomes apparent after about 1 h and is complete after about 6 h.

If a maximal dose of lead is added to a cell suspension, all cells in the population lose potassium at the rapid rate. Under this condition, it is possible to study the mechanism of the lead-induced potassium movements in some detail. It is instructive to follow the uptake of ^{42}K from the medium into the cell: immediately after the addition of lead we find that the radio-potassium enters the cells against its own concentration gradient. The ratio ^{42}K inside/^{42}K medium rapidly increases to values of 2 to 3 but drops thereafter to 1.0. Under the conditions of this experiment, the pump is inhibited. Active transport is not involved. The energy for the transient ^{42}K accumulation is derived from the simultaneous downhill movements of non-radioactive potassium from the cell interior into the medium. We are apparently dealing with a typical example of "counter transport" which is mediated by a carrier system. A more detailed investigation of the described effect suggests that, possibly, lead releases normally inactivated carrier molecules in the membrane and thus indirectly provides a highly specific vehicle for the transfer of potassium across the red blood cell membrane.

The data on which my remarks are based were described in some detail in a recent review (PASSOW, H., "The red blood cell: Penetration, distribution, and toxic actions of heavy metals", in: MANILOFF, J., COLEMAN, J. R. and MILLER, M. W. (Eds.) Effects of Metals on Cells, Subcellular Elements, and Macromolecules, C. H. Thomas Co., Springfield, Illinois (1971), page 291. In this review, further references can be found.

SCHMITT: Thank you very much, Would Prof. SCHLOEGL like to comment?

SCHLOEGL: As a short commentary I want to stress that we are back to the old "antiport" model, in this case competitive transport between K^+ and lead. MITCHELL's idea is that transport in a membrane involves one carrier and two carried substances. We offer the carrier (c) the two substances which can complex with it. Let us assume that substance (a) is at equal concentration on both sides and (b) is present at different levels on the two sides. Then on one side we get the complex predominantly (b—c) and on the other side predominantly (a—c). If (b) differs in concentrations on the two sides, (a), (as a—c), will be transported in the opposite direction. Actually no specific carrier needs to be involved when a specific binding between the membrane substance and the substances (a) and (b) occurs. The membrane is in thermal motion and any complex will move through it according to statistical probability.

SCHMITT: Would Prof. EIGEN care to comment?

EIGEN: I should rather like to present a question to the audience: We have examined a whole series of antibiotics which distinguish well between Na^+ and K^+. All those "carriers" tested so far change their configuration upon ion-binding at a rate of about 10^8 sec-1. The group includes various cyclic polyethers, valinomycin, the actius and others. The kinetics of configurational change always show two steps, which we understand quite well. Most of these carriers are K^+ specific but some are also Na^+ specific, for instance, antamanide, which was characterized by THEO WIELAND and his group. The problem is, what are these substances made for? Nobody knows whether these carriers have any function outside certain microorganisms.

In another meeting, Dr. GREEN from Madison, Wisconsin, reported that he had localized such a substance from mitochondrial membranes. Could anybody reconfirm those findings or has anybody independent evidence for such carrier actions in mammalian membranes?

SCHMITT: Could Dr. LEHNINGER comment on this?

LEHNINGER: The question raised by Prof. EIGEN could be put in another way: are the carriers in eucaryotic cells of high or low molecular weight? After all, apart from the ionophores, most carriers in membranes have very high specificities, higher even than enzymes. Thus the carriers for adenine nucleotides are more specific than the phosphokinases acting on nucleoside phosphates. To me it is difficult to see how low molecular weight carriers can have the necessary specificity for the transport observed in biomembranes. I believe that the high carrier specificity requires a macromolecule, just as the specificity of enzymes requires a macromolecule. The cyclic polypeptide K^+ carriers just described are specific because they form very definite coordination linkages, but such cannot form with most organic metabolites. In the latter cases one deals with binding relationships similar to those between enzymes and substrates. For reasons such as this, we suspect that most membrane carriers are macromolecules.

EIGEN: I am afraid the discussion is heading in a direction which I had hoped to avoid. What you have said is true for many transporting agents, but not for the cyclic peptides which specifically translocate the alkali cations. The feature distinguishing Na^+ from K^+ is their difference in solvation energies. These two ions, spherically symmetrical, differ only little in their diameter. Their specific recognition can come about only through very specific coordination. This is exactly what low molecular weight carriers can provide. One may, of course, also envisage a protein molecule which functions in the same way and, in addition, provides the coupling to the pump mechanism. But in terms of specificity, one cannot do better than with the centrosymmetrical coordinations found in low molecular weight substances, such as valinomycin, etc. We have studied these carriers to clarify their specific properties. The solvation energy of different inorganic cations may vary by as much as 15 kcal thereof, provided the ions are completely desolvated upon complexation; otherwise the energy difference would be inadequate. We should now consider two possibilities: *first*, we can assume a rigid pore which distinguishes Na^+ and K^+ by their size. *Secondly*, we can envisage a carrier mechanism. Which is preferable? Both the rigid pore as well as the carrier must induce the ion to release its hydration shell. This might be difficult since the solvation energy for Na^+ amounts to about 95 kcal/mole and that for K^+ to about 80 kcal/mole. If the entire energy were required for activation, the process would never come to pass. This means that there must be a mechanism which allows the ion to compensate stepwise for the solvation energy.

This might be more easily possible if the carrier itself is partly solvated, so that the complexing process can occur in a continuous fashion, e.g. via a step-by-step substitution of the solvation shell. We have studied such processes. All the ions in question can be desolvated within about 10^{-9} to 10^{-8} sec. Of course, if a cyclic peptide can act in this way, a protein would also potentially be able to do so. In any case, a carrier mechanism provides certain advantages with respect to the desolvation process.

Another question is that of regulation. It may occur through binding or release of carrier by a protein, or both the carrier and the regulatory processes may be facilitated by the same protein.

If we do have a carrier of low molecular weight, we still require a protein for the regulation of the carrier action. Of course, there must be an exchange of charges. A carrier acting on single ions and transporting them in only one direction would otherwise create too large an electric field. The requirements for electroneutrality certainly assure a 1:1 relationship of transport, i.e. either an exchange of one charge class in both directions, or a transport of both positive and negative ions.

Whatever that mechanism is, we require something to distinguish Na^+ from K^+. The question remains whether this distinction is brought about by a rigid pore or a carrier, i.e. a protein or a low-molecular-weight carrier which is controlled by a protein.

LEHNINGER: There have been attempts in both Green's laboratory and our own to isolate from mitochondria small membrane polypeptides having an ionophore or carrier function. In our case we looked for small polypeptides which might contain D-amino acids and GREEN has looked for small peptides soluble in organic solvents. I think there are great experimental difficulties in these approaches because only very small amounts of the postulated ionophores could be anticipated. Moreover, I doubt very much that living cells contain *mobile* carriers; I am more inclined to think that naturally-occurring ionophores or carriers are fixed to the membrane. I would like to call attention to some interesting recent experiments by SELWYN and his colleagues, who have found that trialkyltins can induce chloride-hydroxyl anti-ports, not only in mitochondria and erythrocytes, but also in synthetic phospholipid bilayers. The relationship of the structure of the trialkyltin compounds to chloride and hydroxide transport and their general behavior in lipid phases would make an interesting subject for structural and kinetic investigations.

SCHMITT: While Dr. LEHNINGER is at the microphone, the chairman should invite him to say more about the calcium pump. We now know that the first event in muscle contraction is the entrance of calcium ions from the T-membrane into the myosin-containing fibers. It also seems that the first event in the nerve action wave is not the entry of sodium but of calcium. Dr. LEHNINGER referred to the isolation of a calcium transport system. It would seem that, if this can function in the potential field which triggers the action potential, it must be capable of translocating calcium extremely rapidly and selectively.

LEHNINGER: The mitochondrial Ca^{2+} transport system may play a role in the excitation of muscle, particularly heart muscle. However, the total Ca^{2+} flux required, considered in relationship to the very large number of mitochondria in heart muscle, is really not very great in relationship to either the maximum rate of Ca^{2+} uptake or the total amount that must be sequestered by the mitochondria. Less than 10 nmoles Ca^{2+} per mg mitochondrial protein must be sequestered, less than 1 % of the total *in vitro* capacity of heart mitochondria. Moreover, our evidence strongly suggests that such small amounts of Ca^{2+} are not transported into the matrix but simply bound to the mitochondrial inner membrane in an energy-dependent process which we call membrane-loading. We believe that this Ca^{2+} is released from its binding sites by some kind of conformational change in the mitochondrial membrane following transmission of the depolarization wave from the T-system to the mitochondrial membranes.

SCHMITT: Dr. EIGEN made the point that procaryotes produce ionophores at extremely low concentration, but that we know little about where they reside. FRITZ LIPMANN has recently isolated enzyme assemblies capable of synthesizing decapeptides in procaryotes without RNA templates. These enzymes have to be in a very special array with respect to one another in order to synthesize the decapeptide, and all of this presumably occurs in relation to membranes. To what extent one might have synthesis of peptides without RNA templates in eucaryotes I do not know. Also, we do not know to what extent carriers such as cyclic peptides exist in membranes. This is a challenge to all the biochemists present.

I should like to ask Dr. LEHNINGER if mitochondria were primeval microorganismic invaders; do they perhaps retain in eucaryotes some residual ability to synthesize peptides without RNA templates? Do you think there is any evidence for this?

LEHNINGER: No.

SCHMITT: May I raise another question of specificity? Valinomycin is an excellent and specific potassium carrier in mitochondria, but what localizes this carrier in the mitochondria membrane? I think it has been established by PRESSMAN that valinomycin is a mobile carrier and will penetrate membranes and phospholipid bilayers. Perhaps valinomycin simply acts as a mobile, lipid-soluble ion-carrier. I think nigericin and valinomycin differ in their electrical charges, permitting some specificity. Some of the other cyclic antibiotics may have other specificities, but I am not sure all of them are lipid soluble.

Now let us turn to molecular mechanisms of *cell recognition* and its functioning in *induction* and *differentiation*. Dr. HÄMMERLING talked on this subject, and I should like to ask him to say a little more about surface diversity, patterns of receptors, and so on.

Cell-Cell Recognition in Induction and Differentiation

HÄMMERLING: One of the most astounding questions in nature is how diversity is generated. There are probably no two individuals alike and one must ask whether in any species there is enough genetic information to account for such diversity. For example, in cattle there are eleven blood group systems, each consisting of as many as 300 different alleles, and mathematical tests involving all possible combinations show that the possible phenotypes exceed a number of 2×10^{15}. This according to HILDEMAN is a number larger than all the cattle that have ever lived.

In addition to the variation in the species there is additional diversity within the individual. As Dr. AUERBACH has told us, each cell has its specific place in each organized tissue; not only that, but it also has to be oriented in a very special way. This necessitates active and passive recognition mechanisms which we think must be properties of the surfaces of the cells involved. Therefore, with millions and billions of different cells there is an extraordinary display of diversity. The central question is whether the genome can do the job. The individual has to get along with a certain limited set of genes that specify cell surfaces. In our laboratory we favor a model, originally put forward by Dr. BOYSE, which suggests that diversity can be generated epigenetically.

I should like to summarize a model conceptualized by Dr. BOYSE: if a cell contains two genes *a* and *b* coding for surface components A and B respectively, one would expect three phenotypes, A, B, or AB. But Dr. BOYSE suggests that the surface proximity of A and B could produce a new specificity determined by the overall geometry of the combination. So a fourth phenotype could be A, B, AB, [AB]. This is not totally hypothetical, because there are examples of interaction antigens. These are antigens formed by the combined action of two independent components. E.g. in the B-system of cattle the expression of the two factors B and C creates a third specificity K that is never observed independently from B and C. A review of the literature would reveal further examples of interaction antigens.

One can theoretically extend this to a, b ... n genes, and thus produce a highly diverse surface pattern. Perhaps we should stop worrying about fixed receptors or antigens and consider recognition in terms of pattern complementarities.

SCHMITT: Naturally at each place there would be a greater or lesser binding. Perhaps we could ask Dr. AUERBACH to discuss Dr. HÄMMERLING's point. I think it is pertinent to inquire whether what we actually measure is controlled genetically or epigenetically. I am not sure it matters operationally. In other words, the need for more or fewer genes has so far not been the problem. The question has been: how does variability arise, and also what stabilizes it ? To what extent is it reversible ? Would Dr. AUERBACH comment: What stabilizes surface recognition ?

AUERBACH: Let me give you an example from embryological experiments which, I think, epitomizes the time problem that Dr. FISCHER mentioned. Consider for a moment kidney induction: what induces the formation of tubules ?

Let us place dispersed kidney tissue on one side of a millipore filter and an inducer on the other. We know that one can ordinarily already observe tubules on the kidney side after 24 h and that they are well developed after 72 h. However, if we remove the inducer after 24 h, the tubules develop further; i.e. their differentiation has stabilized. On the other hand, if the tubules are then dissociated with trypsin, they will not reform and the cells act as if they had never been induced. The temporal factor is clear here. First we have cells which cannot make tubules; then we induce them to where they can do this; but when we interfere with their surface structure, we destroy information already established.

With regard to Dr. HÄMMERLING's comments, when and how do A and B get together ? What is the stimulus for this differentiation ? Is it a rapid or slow process ?

I think there must be a whole series of dynamic changes, and we cannot limit our thinking to a single surface change occurring at one particular time which then changes entirely the differentiation of a given cell. We know, for example, that when the ectoderm is induced by mesoderm to become nervous tissue there is a whole sequence of changes which occur, each one causing an alteration in the cell recognition processes as judged by sorting out in reaggregates. Yet when cells are dissociated they reorganize by coaggregating with cells of their own type, hence the phenotypic expression of the cell itself as well as of groups of cells includes surface properties with a high degree of developmental specificity. A key point is to recognize that there are transition phases and that one of the dimensions to be considered in cell surface differentiation is the dimension of time.

SCHMITT: If macerated cerebellar cortex is treated with trypsin and stirred until dissociated into single cells, these will reform, in the culture, the cellular pattern characteristic of the cerebellum. This happens best on the $18\frac{1}{2}$th prenatal day and suggests that surface configurations constituting recognition sites must change constantly during development. In the 1930's the English school believed that highly potent steroid molecules induced specific differentiation in tissue. This approach proved fruitless. It is not so much the molecular effectors but the reacting systems with their recognition mosaics that are important. May I ask Dr. FISCHER if he would say something on this subject ?

FISCHER: I would like to attempt to carry the problem further and to say that the surfaces of cells have diverse receptors — I refer to PAUL

EHRLICH again — that these receptors can appear at different times, that, following binding of a complementary structure something happens, but that complementarity alone does not suffice. Something additional must be added to activate induction and differentiation. This can be well illustrated with immunological methods.

Defined haptens, tied to the receptor by affinity labelling, will not induce differentiation. The hapten must be first linked to a "carrier", and one possible explanation of the necessity of such a carrier is that, besides the receptor area, further areas must be perturbed in order to "trigger" a cell. We must also ask whether triggering is a specific or a non-specific process. Since lymphocytes can be triggered by specific antigens as well as by non-specific agents, e. g. Concanavalin A, lectins and anti-immunoglobulin (directed against recognition sites of a possible immuno-globulin nature), we have three means of "triggering" a cell and can now study the biochemistry involved in each. After a few minutes something happens to permeability, after 4 to 8 h RNA and protein synthesis starts and DNA synthesis follows after 72 h.

We found that phospholipid metabolism is increased after only a few minutes. More fatty acid incorporates itself in lysolecithin, which most probably is generated at a higher rate. Our working hypothesis is that phospholipase A is activated at the triggering site resulting in a local accumulation of lysophosphatides, which might cause local conformational and permeability changes. At present this is merely a working hypothesis. What we do know is that, almost immediately, upon "stimulation" of lymphocytes, increased fatty acid incorporation (but not *de novo* synthesis) takes place. We consider it very important to isolate receptor-bearing areas to study not only the association of complementary structures, but also about the triggering process.

DROEGE: It has been shown in lymphocyte populations that there is a portion of cells which can be triggered by phytoagglutinius and a different portion which can be triggered by anti-immunoglobulin. We are dealing with completely different triggering sites in the two and may eventually have to face two entirely different mechanisms of biochemical triggering. I want to call your attention to the phenomenon of lymphocyte migration, which involves a recognition process. The lymphocyte has to recognize a target organ in order to bring about proper circulation. This is a dynamic process. The recirculating pool of lymphocytes in the organism is originally derived from the thymus and the thymus-derived cells come originally from the bone marrow. There is a cell line of lymphocytes which were originally programmed to migrate to the thymus; later on these cells have to be programmed again to go into the periphery and assume a particular pattern of circulation. That means they are differently programmed for recognition by the target organ. Experiments done by BERTRAND GESSNER demonstrated that the action of sialidases on these lymphocytes changed their circulation pattern to the extent that they could no longer recognize their target organ. But this does not strongly specify the surface components which determine the migration pattern and the recognition of the target organ. However, it leads one to think about the possibility that carrier molecules already present on cell surfaces can be altered, perhaps enzymatically, to bring about new specificities, evoking new recognition processes. The so-called "homing" of lymphocytes is a clear indication of surface recognition. They have to know how to get through the blood vessels to get to lymph nodes. They do so only if they have the right recognition surfaces and can be ruined by enzyme treatment.

Membrane Constitution

SCHMITT: It is only now being realized that the early work by W. J. SCHMIDT on the structural analysis of membranes by optical birefringence still leaves ample room for development. It can be applied to living cells. The method is very sensitive; one can detect, if not measure, the optical interference from red cell ghosts due to a membrane which is only 60 to 80 Å thick, little more than on lipid bilayer. We believe that it is the membrane lipid that accounts for the high birefringence. Most people are not aware of the fact that polarization optics is an extremely sensitive method and, if the extinction coefficient is high enough, it is possible to detect a single lipid bilayer. In the case of the red cell envelope, we have even managed to show the presence of oriented protein in the membrane by means of the difference in form birefringence with solvents of varying refractive index. I hope that these methods, pioneered by INOUE also, will find renewed application in cell biology.

WALLACH: Freeze-etching and freeze-cleaving are powerful new methods of electron microscopy, and we hoped to have Dr. BRANTON here to discuss it. I have merely helped Dr. WEINSTEIN in some of his work. In place of the usual approach, a piece of tissue or block of cells is frozen at a temperature of − 180 °C, split mechanically and the two faces replicated with platinum, gold, or carbon vapor. When viewed in the electron microscope, these replicas do not show the "unit membrane" typical of transmission microscopy, but an appearance which is seen in all biomembranes other than myelin. There is typically an A-face directed outward and a B-face pointing towards the cytoplasm. The face pointing outwards is characterized by a large number of 90 to 100° particles, arranged in clusters and chains. The pointing inward is characterized by few particles. If the nature of the particles is unclear, not so their localization. It used to be thought that the two faces represented the inner and outer surfaces of the living membrane, but it is now clear that the cleavage plane occurs somewhere inside the inner membrane. In some cases, at high resolution one can see particles penetrating. The particle distribution can be altered by phospholipase C; the actual number of particles on the two surfaces does not change but they cluster together, separated by smooth regions. In human erythrocyte membranes, they remain even after substantial trypsin treatment.

SCHMITT: This kind of work is likely to reveal to us the enzyme clusters and the equipment on the surface of the cell which is vital for information processing.

Our own birefringence work started more than 30 years ago, but, as I pointed out, despite its increasing interest, the problems involved are far from solved, and the experts still disagree about the interpretation. Dr. KREUTZ has carried out some interesting studies on this subject, not only on myelin, but also on the membranes of the outer limbs of retinal rods and on chloroplasts. These three represent cases where cell membranes are packed so closely together that one can study them by X-ray diffraction methods. I wonder whether Dr. KREUTZ could briefly summarize the highlights of his studies.

KREUTZ: Prof. LEHNINGER has stressed some points of lipid protein interactions. We find it remarkable that nearly all biomembranes contain $30 \pm 5\%$ lecithin (PC), the one outstanding exception being the photosynthetic membrane, which contains at the most 3% PC, whereas it comprises 23% chlorophylls. On the basis of our X-ray experiments we are led to assume that, in the case of the photosynthetic membrane, the coupling between proteins and lipids is achieved via chlorophyll and, namely, in such a manner that the phytol chains penetrate the protein layer to interact hydrophobically with it while the porphyrin rings contact the fatty acid region of the lipids. There is reason to suggest that in other membranes this coupling action is performed by lecithin. Such a lecithin function is indicated by X-ray results showing that among the phospholipids PC occupies an exceptional position by virtue of its high steric stability. The other phospholipids show an intrinsic dynamic behavior. For instance, phosphatidic acid (PA) and phosphatidylserine (PS) are able to exist in two different conformational states, which are governed by the binding of ligands to their polar group. Such ligands are H^+, Na^+, K^+, Ca^{++} and additionally amino acids in the case of PS (in the one conformational state the polar group just out of the hydrocarbon region, while in the other it is incorporated in it). Phosphatidylethanolamine (PE) is outstanding in that it forms ligand dependent micell types: tubular PE micells are transformed into lamellar micelles by the binding of $HNaCO_3$ or Ca^{++}, etc. PC, however, shows neither the PA- nor the PE-behavior under physiological conditions. It is therefore predestined to a structural role in biomembranes.

Prof. WALLACH has advanced biophysical arguments that biomembranes exhibit transverse and tangential polarity. This means that the evaluation of X-ray diagrams obtained from membrane cross sections should reveal asymmetric electron density distributions and that membrane planes should produce reflections of a two-dimensional lattice (because tangential polarity can only be imagined if it is established by a planar lattice). The X-ray analysis of three biomembranes, namely the photosynthetic membrane, the ROS-disc-membrane (retina rod discs), and nerve myelin indeed reveal asymmetric electron density distributions of the cross section and it has also been possible to demonstrate planar lattices in the photosynthetic membrane and the ROS-disc-membrane.

MENKE: Our discussions, Prof. KREUTZ, have always been concerned with the question whether the X-ray data have an unequivocal interpretation. Other X-ray crystallographers arrive now at symmetric membranes. However, as much as I have tried, I have not succeeded in disproving the Kreutz model of the molecular structure of the thylakoid membrane. Our immunologic data show that proteins (including some which have been highly purified, like ferredoxin, coupling factor and ferrodoxin reductase) lie on the outside of the thylakoid membrane. We also know now, but have not published this yet, that the isolated lamellar system of chloroplasts exists in several states: one, which morphologically looks intact but which functions poorly, and a second state which is morphologically much less intact but which functions very well. These two states differ in their serological properties. When using antibodies to chlorophyll, we find that in the first state of the thylakoid membrane this substance lies as postulated by Prof. KREUTZ, namely under the protein layer. However, we do not know whether there are proteins on the inner side of the thylakoid membrane. Our studies with antisera against thylakoid lipids are also in agreement with Prof. KREUTZ' predictions.

Finally, we believe we have evidence that there is only one protein layer per thylakoid membrane. One of my co-workers has dissolved the membranes in formic acid, removed the lipid and obtained a water-soluble protein of 470,000 Daltons, with a thickness (determined by negative contrast electron microscopy) of 50 Å. Complete thylakoid membranes show the same thickness but contain 40% lipid. If two separate protein layers were contained in the thylakoid membrane, it should be expected that the protein particles were only half the thickness.

SITTE: I am impressed by Prof. KREUTZ' statement that lecithin may penetrate into membrane proteins and that a second lecithin "layer" may be added to this complex, giving a "unit membrane" appearance. But can this be generalized? After all, thylakoid membranes are among the thinnest known. In this connection, let me mention some unpublished work by Prof. KOLLMANN and Dr. KLEINIG. They have isolated a lipid-free protein from sieve tubes of pumpkin. Surprisingly, when this pure protein is combined with vinblastine, the complex exhibits beautiful "unit membranes" in thin sections and a typical biomembrane appearance upon freeze-etching, although no lipids are present at all.

Cooperativity

SCHMITT: We discussed *cooperativity* in membranes a great deal and examined possible ways in which receptors might interact with transducers in the lipid-protein matrix of the membrane. However, the question was raised: How do you know that there is cooperative interaction? Dr. WALLACH wanted evidence in physical terms.

In the process of activation of a receptor there is specific recognition, e.g. of a hormone, present at very low concentration, e.g. 10^{-8} M. If such a receptor is coupled via a transducer to adenyl cyclase, producing cyclic AMP from ATP, a large amplification factor is achieved, producing much higher concentrations of the potent biologic activator, cyclic AMP, than of the hormone that triggered the transduction. One might expect high cooperativity in membranes, not only of the protein components, but also of the lipids, as the recent experiments by TRÄUBLE show.

Transformation

SCHMITT: The word *'transformation'* is not appropriate, since 'transformation' was used by AVERY in a different sense. However, everybody now talks about *transformation* of cells as a result of viral infection, or insult by carcinogenic agents. An important point in this connection is whether cells have preformed surface sites that can combine with lectins like concanavalin A (con. A) and other ligands. MOSCONA has shown that con. A agglutinates embryonic cells that had been dissociated by EDTA. Availability of certain membrane sites seems necessary for differentiation and embryogenesi. MAX BURGER and others have shown that "cryptic receptors" can be uncovered by proteolytic enzymes as well as by infection by oncogenic viruses. The degree to which these sites are functional experimentally depends on their ability to stick to or to react with the con. A. If the sites are exposed, the cells agglutinate upon exposure to the lectins. Contact inhibition is lost when these sites are made available, and such cells display characteristics of malignancy. It is not clear how this works, but membrane changes may play a central role in carcinogenesis. There is direct relation between contact inhibition and the ability to combine with con. A. If you add trypsin to a cell monolayer, the cells lose their contact inhibition and start to divide, piling up in little heaps.

WALLACH: First, I wish to point out that there are plasma membrane defects in transformed cells which go far beyond the appearance of lectin-binding sites. Concerning masked receptors, there are large numbers of chemical sites, some of which are well defined, which can be called "cryptic antigens". The term "cryptic" is used because the receptors become accessible upon mild proteolytic action. Also there are interesting, chemically undefined iso-antigens. The receptor for agglutinin, which BURGER discussed is a different type of receptor. Another one, which is fascinating and which is quite prominent on certain normal cell surfaces, is that receptor for con. A. Finally, SELA and associates found a lectin in soya bean which is quite different from the others in specificity, but also binds to a "cryptic" receptor. The situation can be very complex and I would like to refer to some new findings on that. Firstly, the receptor for the soya bean agglutinin is uncovered in transformed hamster cells. Moreover, it can be uncovered rapidly by protease treatment in normal mouse and human cells, but in a normal hamster cell it appears only after extensive proteolysis. There is thus a striking difference in the receptor accessibility between species. To what extent does this have to do with the neoplastic conversion? Here the recent work of BENJAMIN is quite critical; he has developed a number of polyoma mutants which are *non-converting* neoplastically and with BURGER has examined the appearance of their agglutinability by con. A and wheatgerm agglutinin at various times after virus infection. Within 8 h after infection by such a non-converting virus the agglutination phenomenon appears. So one cannot relate this rigidly to the neoplastic process. Moreover, there certainly are a number of viruses which do not cause tumors, e.g. herpesvirus, which leads to the unmasking of "cryptic" receptors. We are looking at a rebuilding of membrane, occurring very early in viral infections, which is not necessarily related to the transformation conversion process.

SCHMITT: We should not think of the cell membrane as only the inner portion, which is 70 to 100 Å thick, but rather in terms of the "greater membrane" which includes a diffuse outer portion or cell coat. The term "cryptic" does not necessarily refer to the lipid bilayer areas of the inner membrane but may be more relevant to the outer portions that contain glycolipids and glycoproteins. In transformation some process gets triggered that makes hidden groups accessible, and these change the properties of the membrane.

Receptors

SCHMITT: Let us now turn to the problem of *receptors*. I referred to the isolation and purification of receptors, particularly the recent work of MILEDI, MOLINOFF, and POTTER, using specific binding proporties of snake venom toxin. Possibly in a few years, a whole series of receptors will be isolated and their subunit structure determined. Two people who have worked in this general area have asked to make brief presentations. Dr. WESEMANN and Dr. MARKS have obtained information about the 5-OH tryptamine and sialic acid in a membrane that is particularly precious to me, the synaptic membrane. Dr. WESEMANN, please:

WESEMANN and MARX: During the last two days we have learnt much about different aspects of membrane function: about the metabolic traffic across the membranes of mitochondria, about the regulation of cell replication and cell association by surface membranes and about the expression of surface antigens. I appreciate that the chairman gives me the opportunity to make a few comments on another type of membranes with a highly specialized function, i.e. about the nerve-ending membrane and their importance for the transmission of nerve impulses. According to the concept of neurotransmission, the transmitter substances are stored in the vesicles of the nerve terminal and released into the synaptic cleft upon stimulation where they react with the subsynaptic membrane as the target tissue for bioelectric alterations. Thus three membrane structures are essential for neurotransmission: the membranes of the synaptic vesicles as the storage sites of the transmitter molecules, the presynaptic and the subsynaptic membranes where the receptors for the binding of the transmitter must be located.

In order to characterize the receptor molecules and to elucidate the functional interrelationship between chemical and morphological structure, we investigated the storage and binding of the transmitter substance

5-hydroxytryptamine in synaptic membranes and vesicles isolated from rat brain.

5-hydroxytryptamine was extracted with butanol/water from vesicles containing 0.2 nMol 5-hydroxytryptamine/mg protein and found to be enriched in the butanol phase. After extraction of the nerve-ending membranes, on a protein basis, 5-hydroxytryptamine as well as N-acetyl-neuraminic acid were concentrated in the butanol layer. The enrichment of 5-hydroxytryptamine in the organic phase, which is in contrast to the partition of unbound 5-hydroxytryptamine, could mean that the distribution of 5-hydroxytryptamine is mediated by a carrier or receptor substance. The high relative specific concentration of N-acetyl-neuraminic acid and the absence of gangliosides in the butanol phase support the hypothesis that this carrier substance can be classified as sialoglycoprotein or sialoglycolipid. This would be in accordance with results obtained by FISZER and DEROBERTIS who isolated from nerve-ending membranes a "5-hydroxytryptamine-complex" with the chromatographic properties of a proteolipid. The distribution pattern of sialic acid and 5-hydroxytryptamine between butanol and water suggests that this carrier substance is similar in composition to the 5-hydroxytryptamine receptor of smooth muscle, where a correlation was found between 5-hydroxytryptamine binding and sialic acid metabolism.

As a morphological equivalent, neuraminic acid-containing substances could be demonstrated by electron microscopy. Electron-dense material was precipitated on the membranes after treatment with colloidal iron hydroxide (Hale's stain) marking the sites of reaction between iron hydroxide and anionic groups. At the low pH of 1.7 at which the reaction with the nerve ending membranes was carried out, only N-acetyl-neuraminic acid and monosulfates are dissociated enough to precipitate iron hydroxide. To differentiate between these two reactive groups, sulfate was split off by transesterification with methanol while sialic acid was removed enzymatically. After incubation with neuraminidase [E. C. 3.2.1.18], which released about 95% of the N-acetyl-neuraminic acid bound to the membranes, only the sulfated groups are stained with iron hydroxide. The specificity of our staining procedure for acidic groups was proved by the complete absence of iron granules after esterification with iron hydroxide. After saponification the reactivity towards iron hydroxide was only partly restored. The iron granules indicate the presence of sialic acid in the membranes, since sulfated mucopolysaccharides are desulfated by methanol.

To sum up our results: sialic acid-containing structures are constituents of nerve-ending membranes, as could be demonstrated by the Hale stain reaction. The chromatographic properties and the phase distribution of this compound and of 5-hydroxytryptamine are in accordance with the concept that it can be characterized as a proteolipid which may act as receptor for the transmitter 5-hydroxytryptamine.

SCHMITT: In the case of bacteria, we know that RNA and DNA are bonded to the membrane. In animal cells the genetic material is primarily in the cell center in the nucleus. One thing that came up clearly throughout the colloquium is that membranes contain important switching equipment of the cell, and part of this has to do with a programmed turning on and off of genes. I have pointed out that when you stimulate the postsynaptic element of a neuron, you activate the DNA to produce the type of messenger RNA required to synthesize enzymes necessary to produce transmitter substances. The Neurosciences Research Program recently held a work session on this subject and found a whole series of reactions that take place at the cell membrane and trigger changes in the nucleus. Nobody quite understands how information is transmitted from the cell surface to the nucleus. Information also flows in the reciprocal direction by the action of gene products on the cell membrane. The equipment for such information transfer must exist in the cells. We have known about microtubules for only ten years, but it seems probable that microtubules may be intimately concerned with the rapid and directional transport of substances in cells. In nerves, material generated in the neuron cell body is transported down the axon at high velocities (200—2000 mm per day). I expect that the next few years will bring substantive revelations about the mechanism of this transport.

FISCHER: OTTO WIELAND, the president of our society, has asked me to close the Round Table Discussion and the 22nd Mosbach Colloquium. All I have to say is that to my mind the meeting was effective in that it formed a bond — as tight as a membrane — between lecturers and the audience. I hope that with the help of the contributors and discussants we will be able to put together shortly a small readable book which will reflect the stimulating atmosphere of this meeting. Speaking of atmosphere and high spirits I have to thank all of you, the silent majority, the discussants, the lecturers and last but not least Professor FRANCIS O. SCHMITT, who truly was the king of Mosbach!

I hereby close the 22nd Mosbach Colloquium and look forward to the 23rd. Auf Wiedersehn and thank you!